Annual Conference of the American Forage and Grassland Council (AFGC 2025)

Kissimmee, Florida, USA
12 - 15 January 2025

ISBN: 979-8-3313-1337-1

Printed from e-media with permission by:

Curran Associates, Inc.
57 Morehouse Lane
Red Hook, NY 12571

Printed by Curran Associates, Inc. (2025)

For permission requests, please contact American Forage and Grassland Council (AFGC) at the address below.

American Forage and Grassland Council (AFGC)
PO Box 867
Berea, KY 40403
USA

Phone: 800.944.2342

www.afgc.org

Additional copies of this publication are available from:

Curran Associates, Inc.
57 Morehouse Lane
Red Hook, NY 12571 USA
Phone: 845-758-0400
Fax: 845-758-2633
Email: curran@proceedings.com
Web: www.proceedings.com

TABLE OF CONTENTS

MONDAY POSTERS

ORAL PRESENTATIONS

TUESDAY POSTERS

FORAGE TRENDS IN THE SOUTHEASTERN HAY CONTEST

L. Baxter[1], U. Saha[2], K. Payne[3], and L. da Silva[4]

The Southeastern Hay Contest (SEHC) recently celebrated 20 years of proudly recognizing producers who grow and harvest high quality hay. The contest was originally developed by local Extension agents to bring awareness to the importance of hay testing and managing livestock feed needs through nutritive value determination. Extension agents and specialists not only promote the contest, but also use it in education and outreach throughout their counties and regions. Over time, the contest has expanded to nine categories that producers can enter forage samples, including: warm-season perennial grass hay, warm-season annual grass hay, alfalfa hay, all other legume hay, cool-season perennial grass hay, grass-legume mixed hay, cool-season annual grass hay, grass baleage, and legume baleage. The samples are ranked based on relative forage quality (RFQ) before winners are determined. The contest has grown steadily over time and a record number of entries was reached in 2024 with 496 submissions from nine states across the region. These data provide unique insights to forage production trends across the region. For instance, a greater number of samples tend to be disqualified for high moisture in heavy rainfall years. Also, it is not uncommon for elevated reports of nitrates during periods of drought. Summation of the data from the SEHC can provide an interesting reflection on forage trends over time.

[1]Assistant Professor and Extension Forage Specialist, Crop and Soil Science Department (baxterl@uga.edu); University of Georgia Tifton Campus, Tifton, GA. [2]Program Coordinator, Feed & Environmental Water Laboratory (sahau@uga.edu); University of Georgia, Athens, GA. [3]Applied Forage Systems Specialist (kathrynpayne@vt.edu); Virginia Tech, Southern Piedmont Agriculture Research and Education Center, Blackstone, VA. [4]Assistant Professor and Extension Forage Specialist, Animal and Veterinary Sciences Department (lseveri@clemson.edu); Edisto Research and Education Center, Blackville, SC.

HOW DOES NIRS WORK?

Jasmine Major[1], Lisa Baxter[2], William Fleming[3], and Justin Burt[4]

Near infrared spectroscopy (NIRS) is a powerful tool widely used by producers and researchers across forage and livestock industries to predict forage nutritive value. This tool can provide rapid and effective measurements of forage nutritive value parameters and has been widely adopted in commercial, private, and research laboratories across the United States. Unfortunately, many people do not know how this technology works or how results are generated for their forage report that returns from the laboratory. The objective of this poster was to briefly outline the fundamentals of NIRS that are most relevant to forage-livestock producers and applied scientists. We will illustrate how NIR light interacts with a forage sample to produce spectral characteristics. Next, we will discuss how calibration are applied to these characteristics and forage quality parameters may be predicted to be reported for forage reports. We will briefly discuss limitations of NIRS that must be observed in the interpretation of the results. This poster will specifically focus on calibrations developed by the NIRS Consortium (NIRSC). The NIRSC is a collaboration of commercial laboratories, universities, government groups, plant research entities, and instrument companies. The mission of the NIRSC is to promote and standardize the use of NIRS through the development of robust, accurate prediction calibrations.

[1]Graduate Research Assistant, Crop and Soil Science Department (jasmine.major@uga.edu); University of Georgia Tifton Campus, Tifton, GA. [2]Assistant Professor and Extension Forage Specialist, Crop and Soil Science Department (baxterl@uga.edu); University of Georgia Tifton Campus, Tifton, GA. [3]Graduate Research Assistant,, Crop and Soil Science Department (william.fleming@uga.edu); University of Georgia Tifton Campus, Tifton, GA. [4]Postdoctoral Research Associate, Crop and Soil Science Department (justin.burt@uga.edu); University of Georgia Tifton Campus, Tifton, GA.

COUNTY AGENT PARTICIPATION INFORMS A BAHIAGRASS TRIAL

J.O.C. Kubesch and K.J. Simon[1]

Long-term variety trials can be a challenge for new specialists with 100% Extension appointments. Agents and farmers often frame the research questions that drive applied forage programs. However, these stakeholders often miss out on experiential learning opportunities that shape the development of in-field trials. This project sought to include county agents in a pilot trial that will shape a long-term field trial in southwestern Arkansas. Alongside a regional trial conducted in southwestern Arkansas, a controlled trial was assembled at a site in Saline County, Arkansas. Buckets of topsoil layered over potting media were planted to one of three commercially available bahiagrass varieties as well as a common bermudagrass. Half of the buckets were assigned to common producer practices (CPP) as described by a panel of 15 county agents, and the others were assigned to best management practices (BMP) per existing Extension recommendations. Canopy height and forage emergence were followed throughout the growing season. Canopy height did not differ among stands. Stands managed to BMP had more sown cover than stands managed to CPP. Establishment success was categorized as fully, moderately, or partially functional. Bahiagrass established slightly more successfully than common bermudagrass. The trial only produced a single harvest, 23 weeks after planting. Armyworm pressure was high in the late summer and reduced the harvest. This trial and the regional trial informed expectations for a first year stand as well as setting management expectations for a field trial, slated for planting in spring 2025.

[1]Forage Specialist, University of Arkansas System Division of Agriculture, Cooperative Extension Service, Little Rock, AR (jkubesch@uada.edu); Forages Instructor, University of Arkansas System Division of Agriculture, Cooperative Extension Service, Little Rock, AR (ksimon@uada.edu).

RHIZOMA PERENNIAL PEANUT IN CENTRAL ALABAMA

J.B. Elmore, M.D. Price, L.S. Silva, S.L. Dillard and M.K. Mullenix[1]

Rhizoma perennial peanut (RPP, *Arachis glabrata*) is a perennial forage legume with high yield and nutritive value characteristics. It is well adapted to the southern United States and can be managed under hay production or grazing management, including in mixtures with bahiagrass (Paspalum notatum). While RPP has been growing in south Alabama for decades, this demonstration was conducted to evaluate RPP's persistence in central Alabama. The Chilton Research and Extension Center is located on Upper Coastal Plain soils near the geographic center of the state. Prior to planting, strips were sprayed using glyphosate to kill existing vegetation in a bahiagrass sod and lime and fertilizer were applied according to soil test recommendations. The strips were tilled in preparation for planting. On July 9,2021 three varieties of RPP, UF Tito, UF Peace and Florigraze were planted within the tilled strips at the recommended planting rate and cultipacked. Leaf emergence occurred on July 26, 2021, with the first bloom appearing on August 2, 2021. Since 2021, all three varieties have persisted and survived flash droughts in the summers and extremely cold temperatures in the winters. Plots have been maintained by mowing and spraying when deemed necessary. Although all three of the plantings have persisted, UF Peace and UF Tito are the most productive in this demonstration.

[1]Animal Science and Forage Regional Extension Agent, Alabama Cooperative Extension System, Clanton, AL (elmorjb@auburn.edu); Director, Chilton Research and Extension Center, Auburn University, Clanton, AL (pricemd@auburn.edu); Assistant Professor/Extension Specialist, Clemson University Edisto, SC (lseveri@clemson.edu); Associate Professor/Extension Specialist, Auburn University/Alabama Cooperative Extension System, Auburn, AL (dillasa@auburn.edu); Department Head and Professor Animal Sciences, Auburn University, Auburn, AL (mullemk@auburn.edu).

HEIFER DEVELOPMENT IN SOUTHSIDE VIRGINIA

K. Payne, T. Clarke, M. Henry, J. Currin, and C. Gregg[1]

Maintaining a productive and profitable beef cattle operation is integral for the longevity of its success. As part of this success, it is important to ensure that young heifers are capable of producing healthy calves. First time heifers often require a separate, more focused feeding program and care in navigating their first breeding season rather than remaining in the herd with more mature cows. In 2021, a collaborative effort among the Virginia Tech Southern Piedmont AREC, Virginia-Maryland School of Veterinary Medicine, local Virginia Cooperative Extension Agents, and Amelia Area Cattlemen, LLC., established the heifer development program at the Southern Piedmont AREC in Blackstone, VA. Heifers from local producers arrive at SPAREC in September where they are evaluated to pass the Virginia Premium Assured Heifer standard. The heifers are rotationally grazed on stockpiled novel tall fescue pasture throughout the fall and winter and are fed grain daily to ensure body condition by artificial insemination breeding in December. Heifers remain with bull exposure for 60 days and return to producers after pregnancy checks in early March. Forage yield and nutritive value of the stockpiled tall fescue are measured throughout the duration of the program each year. Over the four years of the heifer development program, participation has increased from 42 heifers from four producers in 2021 to 65 heifers from six producers in 2024. Pregnancy rates have remained consistent between 90 to 95% in the first three years, with a range of 45 to 62% rate from AI.

[1]Applied Forage Systems Specialist, Virginia Tech Southern Piedmont AREC, Blackstone, VA (kathrynpayne@vt.edu); ANR Agent – Mecklenburg County, Virginia Cooperative Extension, Boydton, VA (cclarke@vt.edu); Amelia Area Cattlemen, LLC., Amelia, VA (mike.henry123@verizon.net); DVM and Clinical Associate Professor, Virginia-Maryland College of Veterinary Medicine, Blacksburg, VA (jcurrin@vt.edu); ANR Agent – Brunswick County, Virginia Cooperative Extension, Lawrenceville, VA (clgregg@vt.edu).

DEVELOPMENT OF CLIMATE-SMART EDUCATIONAL EFFORTS FOR FORAGE AND LIVESTOCK PRODUCERS IN SOUTH CAROLINA

L. Silva, M. Miller, H. Malcomson, R. vanVlake, C. Garcia, L. Vieira, R. Noble, N. Mowa, K. Flynn, C. Kurz, and P. Agudelo[1]

Abstract

The livestock industry is one of the most important agricultural activities in the Southeast. Over the years, educational efforts have focused on improved management practices to optimize forage and animal production to reduce production costs due to the fluctuations of input prices. In recent years, there has been increasing interest in sustainable production intensification on forage-livestock ecosystems aiming to reduce the environmental footprint of operations. Despite the popularity that terms such as 'climate-smart practices' have gained, research-based educational resources for Extension education, agricultural professionals and producers are still needed. In 2022, the United States Department of Agriculture announced the "Partnerships for climate-smart commodities" which funded over 140 projects and invested $3.1 billion. In collaboration with South Carolina State University, Clemson University was awarded $70 million for a project that addresses four agricultural commodities: forages for beef cattle, peanuts, forest and leafy greens. Over 150 livestock producers are enrolled statewide, and greenhouse gas measurements from soil and animals and soil sampling will be conducted. This pilot project provides an annual stipend for producers to implement selected improved management practices (incorporation of clovers, poultry litter and improved grazing management). In the first year, the project offered 15 official program training and outreach events reaching over 600 forage and livestock producers through in-person events. Feedback provided by producers has allowed educators to adjust educational strategies and create additional resources to facilitate the understanding of concepts and facilitate and expand the adoption of improved management practices.

Introduction: The livestock industry is one of the most important agricultural activities in the Southeast (NASS Census, 2017). Forages are the primary feed source used by livestock operations due to favorable climatic conditions, a wide range of adapted species, and access to local resources (i.e., poultry litter) in the region. Under grazing management, forages also contribute to lower production costs once the need for harvesting, storing, and transporting feed decreases. In the Southeast, most forage systems are based on the use of perennial grasses such

---

[1] Assistant Professor, Clemson University, Blackville, SC (lseveri@clemson.edu); Extension Associate, Clemson University, Clemson, SC (mmille6@clemson.edu); Graduate student, Clemson University, Clemson, SC (hmalcom@clemson.edu); Senior Extension Agent, Clemson University Cooperative Extension System, Florence, SC (rii@clemson.edu); Postdoctoral Fellow, Clemson University, Blackville, SC (cvelaga@clemson.edu); Postdoctoral Fellow, Clemson University, Blackville, SC (lvieira@clemson.edu); CAFLS Dean of Research, South Carolina State University, Orangeburg, SC (rnoble3@scsu.edu); Project Manager, South Carolina State University, Orangeburg, SC (cmowa@scsu.edu); Project Manager, Clemson, SC

(kgilker@clemson.edu); CAFLS Dean of Research, Clemson University, Clemson, SC (pagudel@clemson.edu).

as bahiagrass (*Paspalum notatum* Flügge), bermudagrass (*Cynodon dactylon* L.), and tall fescue (*Festuca arundinacea* L.). Grasses rely on nitrogen (N) fertilizer inputs to sustain forage production and support animal performance (Silveira et al., 2014), elevating the activity's carbon footprint. The seasonality of the production of perennial grasses is a major challenge for livestock operations, which requires the use of supplemental feed, primarily hay (Redmon, 2000), or the incorporation of management practices (e.g., cool-season forages, stockpiling) to extend the grazing season. Over the years, educational efforts have focused on improved management practices to optimize forage and animal production to reduce production costs due to fluctuation of input prices. In recent years, there has been an increased interest in developing and incorporating technologies to support the reduction of greenhouse gas emissions and off-farm inputs while enhancing nutrient cycling and carbon sequestration in forage-livestock ecosystems. Therefore, educational efforts should be designed to meet contemporary needs and promote research-based recommendations and adoption of technologies by producers.

According to the USDA (USDA, 2023), climate-smart practices are practices that can improve the ecological, social, and economic resilience of agricultural systems to climate change, enhance carbon sequestration, reduce greenhouse gas emissions, and improve soil health. For example, climate-smart practices include the adoption of improved grazing strategies and ruminant feed management. Over the last couple of decades, there has been an increasing interest in forage research to better understand practices that improve carbon and nutrient accumulation in the soils under forage-based systems. Soils under perennial grasslands are essential sinks of soil organic carbon (SOC) (Conant et al., 2017) due to continuous plant biomass input and minimal soil disturbance (Cambardella and Elliott, 1992; Six et al., 1998). Implementation of new technologies and the use of climate-smart practices will continue to grow and be essential to enhance sustainability and improve the use of resources in forage systems. Providing educational materials and training on those topics is essential, given the lack of research-based resources and the misinformation available in general media outlets. Agricultural educators and farmers have an important role in helping to support reliable information and technologies to be adopted into agricultural systems for the long-term improvement an resilience of forage and livestock ecosystems.

Material and Methods: In the first year of implementation of the project, a total of 15 official educational forage and livestock-based events were planned and delivered. These educational programming activities addressed specific content required and provided guidelines for implementation and management of practices by enrolled producers. The events were also open to the community and local producers could attend them to learn more about the content being delivered as well. Over 600 forage and livestock producers attended the in-person events which consisted of field days, workshops and trainings to producers and educators. As part of the activities promoted and delivered an in-service training was held for the Clemson University Cooperative Extension System Livestock and Forages team in March 2024. During this training, Extension agents learned more about the establishment, management, and nutrition of several forage systems and specific questions related to the climate-smart project were also addressed. Dr. Silva developed and delivered a demonstration guide for the Extension agents with detailed explanations about different demonstration activities they can integrate in their programs and explained the activities during the in-service training as well.

Results and discussion: Surveys were developed and applied to participants post-training. Over 70% of participants answered they would adopt management practices discussed during events in the next 12 months. Among the topics addressed, participants wanted further information and training on soil health, incorporation of legumes, and weed control on forage systems as the top three priority topics. This is aligned with the increasing interest in sustainable production intensification and incorporation of improved management practices by producers. Over the past years, consumer perceptions on how food is produced, and the industry's carbon footprint has been growing, which has also enhanced the need to better understand sustainability and resilience strategies and technologies in order to be adopted into operations. All participants appreciated the combination of classroom concepts and content with outside, practical, and hands-on activities. This approach is essential to exemplify and consolidate main concepts and help participants to visualize some points by seeing plants, working with measuring tools, or seeing the impact of a certain management practice implemented in a system and being able to compare if with a "business as usual" area/field. The training events were held statewide and achieved a diverse group of participants with varying levels of experience in the forage and livestock industry. This was crucial to enhance the exchange of knowledge and practical experiences among participants, and promote networking and fellowship among them.

Literature Cited:

1.Cambardella, et al. 1992. DOI:10.1002/eap.1473
2.Conant et al. 2017. DOI: 10.1002/eap.1473
3.NASS. 2017. https://www.nass.usda.gov/Publications/AgCensus/2017/
4.Redmon, L.A. 2000. Proc of the Annual Meeting Southern ASAS. p. 6
5.Silveira, et al. 2014. DOI: 10.2134/FG-2013-0041-RV
6.Six et al. 1999. DOI:10.2136/sssaj1999.6351350x

Figure 1. Climate-smart educational programming activities delivery: (left) workshop for beginner farmers, (central) in-service training for Extension agents, and (right) field day on adoption of improved management practices and climate-smart practices implementation.

MAINTAINING A FORAGE VARIETY TESTING PROGRAM IN THE 21ST CENTURY

Gene Olson[1], S. Ray Smith, Brittany Hendrix, Chris Teutsch, and Jimmy Henning

University forage variety testing has declined in many states over the last 25 years because of reduced funding and other factors. At the Univ. of Kentucky (UK) we have been able to continue a strong program for many reasons including: 1) maintaining hard funding for coordinators salary, 2) maintaining close connections with industry, 3) multiple forage species, 4) multiple testing protocols (hay, grazing, quality), 5) producer friendly reports, and 6) county agent and producer interest and support. The UK Dept. of Plant and Soil Sciences has been committed to providing hard funding for not only the forage variety test coordinator, but also for corn, soybeans, and small grains. This secure funding has ensured dedicated and experienced personnel and keep variety entry fees reasonable. Permanent funding has also allowed the inclusion of varieties not officially entered by companies, but which are important to KY producers. The forage variety test coordinator and faculty supervisor have worked diligently to maintain close connections with the seed industry. These efforts foster collaboration and provide continued support for the program. Companies have the additional benefit of hosting mini-field days at the UK farm with regional marketers and contracting private trials. Testing all major forage species for multiple traits has helped to maintain relevance of the program. A very successful effort was the development of the annual Forage Variety Summary Report. This publication is a user-friendly guide for choosing the best variety based on a summary of forage yield and grazing tolerance trials conducted in KY over the past twenty years. To access all UK reports go to https://forages.ca.uky.edu and click on the link for "Forage Variety Trial".

[1]Gene Olson, University of Kentucky (raysmith1@uky.edu); remaining authors Univ. of KY.

BAHIA BUCKET TRIALS BENEFIT BEEF AND FORAGE PRODUCERS IN ARKANSAS

S. Hill, J.O.C. Kubesch, and K.J. Simon[1]

Bahiagrass is a prominent warm-season perennial forage in southwestern Arkansas. It has a reputation for being a low-input alternative to bermudagrass. Most bahiagrass acreage in Arkansas is considered "Pensacola". Variety trials in other states have led to interest in using improved varieties, such as "Tifton-9" and "TifQuik". Previous specialists have questioned whether the yield advantages seen in improved varieties will translate under common producer practices. This demonstration project sought to evaluate three varieties of bahiagrass under best management practices and common producer practices. Buckets of field soil were sown to one of the three varieties in May 2024 and then grown out with rainfall. A one-time watering event occurred in August 2024 to offset excessively dry media during a fall drought. Stands were monitored for canopy height, emergence, and success weekly. The trial was a blind study; agents did not know which bucket was assigned to which variety. Agents reported weekly to the forage specialist, and regularly discussed trial logistics, weather, and weed concerns. Fifteen counties across the Ouachita district were involved initially, however only eight reported regularly. Stands were harvested in July and October. The average dry matter yield for the season was 3000 lbs/acre. The trial suggested that bahiagrass plantings in Arkansas can be ready for harvest ~60 days after planting and produce two modest harvests. Agents were able to transport buckets to meetings for education and outreach. A field trial will follow in spring 2025.

[1]Calhoun County Staff Chair, University of Arkansas System Division of Agriculture, Cooperative Extension Service, Hampton, AR (shill@uada.edu); Forage Specialist, University of Arkansas System Division of Agriculture, Cooperative Extension Service, Little Rock, AR (jkubesch@uada.edu); Forages Instructor, University of Arkansas System Division of Agriculture, Cooperative Extension Service, Little Rock, AR (ksimon@uada.edu).

EVALUATING PROMISING PINTOI PEANUT LINES IN NORTH FLORIDA

F. Rondino, C. Mendes, E. Vieira, G. Abitante, R. Trumpp, and C. De Souza[1]

Arachis pintoi, commonly known as Pintoi peanut, is a perennial peanut cultivated as a cover crop and pasture in tropical regions. Although it demonstrates tolerance to acidic soils, heavy grazing, and poorly drained soils, as well as persistence, its use is not widespread in Florida due to the lack of adapted varieties. Unlike rhizoma peanut, which reproduces through rhizomes, Pintoi peanut reproduces by stolons and has the potential to be propagated by seeds, making it a valuable option for integrating with warm-season perennial grasses. It enhances forage systems by fixing nitrogen biologically and improving nutritive value. This research evaluates 19 advanced Pintoi peanut lines alongside the check variety Golden Glory at the North Florida Research and Education Center in Marianna, using germplasm sourced from Dr. Kretschmer's collection at the Indian River Research and Education Center. Established in 2019 on Orangeburg loamy sand soil in a randomized block design with four replications, the plots have endured four winters with minimal management. In 2024, rainfed plots were clipped to a 1-inch stubble every six weeks to assess herbage mass accumulation, canopy height, and ground cover. Results revealed significant genotype differences and genotype-by-harvest interactions. Average forage accumulation per harvest was 800 kg/ha, with top-performing genotypes exceeding 1000 kg/ha, outperforming Golden Glory. Seed production and root biomass evaluations are ongoing, with a second year of assessment planned.

[1]University of Florida, North Florida Research and Educational Center, Marianna, Florida (a.fioccorondino@ufl.edu).

GENETIC DIVERSITY IN SEED RETENTION OF BAHIAGRASS (*PASPALUM NOTATUM*)

A. Valentini[1], J. Longhi[2], A.P. Brunes[3], C.H. de Souza[4], B. B. Vigna[5], H. Mendonça[6] and E. Rios[7]

Bahiagrass *(Paspalum notatum)* is one of the most important grasses in southern Brazil, with significant potential for genetic improvement. However, seed production has been limited due to significant losses from seed shattering. This study aimed to conduct a genetic characterization of a Bahiagrass population in terms of seed retention strength (SRS), to identify genotypes with superior SRS. The experiment was conducted in a greenhouse, with 234 ecotypes from different locations across the Americas, using a randomized complete block design, with three replications. Seed retention was evaluated through the quantitative determination of tensile strength, in grams (g), using a digital force gauge on three seeds from the central region of the raceme. Each inflorescence was evaluated at three different times: 14, 21, and 28 days after anthesis. Other traits such as raceme length (RL) and number of seeds per raceme (NSR) were also evaluated. The genotypes showed higher retention forces at 14 days and lower ones at 28 days. SRS varied from 6 to 66 g and the calculated heritability of the trait was 53%. There is a negative correlation between SR and RL and a positive correlation between SR and NSR. Our data showed that it is possible to obtain genetic gains to increase the seed retention over the next generations, however selecting genotypes with higher seed retention strength would decrease the raceme length.

[1,4]University of Florida (UF), IFAS – North Florida Research and Education Center (NFREC), Marianna, FL (arthur.valentini@ufl.edu), (c.lopesdesouza@ufl.edu). [2,3]Federal University of Rio Grande do Sul (UFRGS), Porto Alegre, RS, Brazil (julialonghi98@gmail.com), (andre.brunes@ufrgs.br). [5]EMBRAPA Pecuária Leste, São Carlos, SP, Brazil (bianca.vigna@embrapa.br). [6]Universidade de São Paulo (USP), São Paulo, SP, Brazil (heitorcalux@usp.br). [7]University of Florida (UF), Agronomy Department, Gainesville, FL (estebanrios@ufl.edu).

SELECTION AND RECOMMENDATION OF UROCHLOA SPP. HYBRIDS BASED ON SEED YIELD USING BAYESIAN AMMI

C.T.E. Mendes[1], C.H.L. De Souza[1], S. Hoppen[2], R. Sampaio[2], T. Madeira[2], A. Silva[2], C.P. da Silva[3] and L.A. de Oliveira[4]

Urochloa spp. (also known as Brachiaria) is a key forage for livestock in tropical regions and has been introduced to subtropical areas like Florida. Mulato II, an interspecific hybrid, performs well on South Florida's well-drained soils, providing high forage quality and yield. However, tropical grasses in this genus typically exhibit poor seed production, leading to higher seed costs for producers. Our study evaluated the stability and adaptability of seed yield in palisadegrass hybrids (*Urochloa spp.*) in Brazil across three locations, two seeding dates, and two years (2022/23 and 2023/24). Data analysis employed the additive main effects and multiplicative interaction model (AMMI) using a Bayesian approach. The multi-environment dataset comprised 14 pre-commercial hybrids, including Camello and Cayman, along with the checks Mulato II, Ruziziensis, and Marandu. Seed yield, measured in kilograms per hectare, was analyzed in a randomized complete block design with four replications. The results indicated that genotypes with higher seed yields contributed significantly to the genotype-by-environment interaction (GEI). Six homogeneous subgroups of environments were identified, suggesting that future selections could rely on fewer locations to evaluate seed yield performance in elite hybrids. Genotypes G8, the check Ruziziensis, and G7 demonstrated outstanding seed yield across all locations. Selecting forage varieties with high seed yield is crucial to ensuring affordable seeds for producers, ultimately lowering the cost of establishing new pastures.

[1]North Florida Research and Education Center (NFREC), U.S., University of Florida, Marianna, FL; [2]Papalotla Seeds, São Paulo, Brazil; [3]Universidade Federal de Lavras (UFLA), Lavras, Brazil, [4] Universidade Federal da Grande Dourados (UFGD), Dourados, Brazil.

THE SOUTHERN COVER CROP VARIETY TRIAL: ASSESSING FORAGE QUALITY OF COVER CROP SPECIES/VARIETIES OVER A 10-STATE REGION

V.R. Sykes[1], A. Gamble[2], A. McWhirt[3], C. Chase[4], G. Maltais-Landry4[d], L. Sharma[4], D. Treadwell[4], N. Basinger[5], R. Hitchcock[5], E. Haramoto[6], C. Reberg-Horton[7], S. Narayanan[8], N. Rajan[9], and M. Reiter[10]

The Southern Cover Crop Variety Trial began in fall of 2022 in collaboration with the Southern Cover Crops Council and covers a 10-state region of the Southern US (TN, FL, NC, KY, VA, SC, TX, GA, AR, and AL). The goal of this program is to provide industry, breeders, and growers with important information on regional adaptation of cover crop species and varieties across a broad geographic area. This trial is fee-based, and results are shared publicly every year. Trial sites are planted in mid-October. Canopy cover, weed suppression, height, and biomass are measured in early and late spring, corresponding to two weeks prior to corn and soybean planting dates, respectively. In years one and two, select states (FL, GA, KY, NC, TN) also measured forage quality (CP, IVTDMD48, ADF, NDF, and lignin) and estimated nitrogen release. Twenty entries each were evaluated in the 22-23 and 23-24. Performance varied significantly among and within locations, varieties, and evaluation period each year. Cereals were top-performers for biomass across most regions both years. Legumes showed greater regional variation, emphasizing the importance of these trials in identifying the right species/variety for a specific region and system. Generally, crude protein declined while IVTDMD48 increased in the early vs. late spring evaluations across all varieties. While balansa, Persian, and red clover tended to have the highest IVTDMD48, they also produced some of the lowest biomass. Full results can be found on *search.utcrops.com*.

[1]University of Tennessee, [2]Auburn University, [3]University of Arkansas, [4]University of Florida, [5]University of Georgia, [6]University of Kentucky, [7]North Carolina State University, [8]Clemson University, [9]Texas A&M University, [10]Virginia Tech.

ASSESSING CARBON MINERALIZATION RATE ACROSS DIFFERENT LAND USES IN FLORIDA

C.E. Souza de Sousa[1], J.C. Dubeux Jr.[2], I.L. Bretas[2], L.M.D. Queiroz[2], L. Garcia[2], M. Ruiz-Moreno[2], B.E. Bizzuti[3], and M.A. Lira Jr.[4]

Soil carbon emissions are associated with soil biological dynamics and can be used to assess soil health and the capacity of microbial activity to decompose crop residues and stimulate nutrient cycling. This study evaluated different land-use systems on CO_2 emissions from carbon mineralization and biological activity. Soil samples were collected from the 0- to 6-cm surface layer on 20 farms located in the state of Florida. Land uses included cropland, pastureland, hay field, low-input grassland, and native area. The collected soil was rehydrated and incubated for 24 days. There was no difference in C-CO_2 emissions between pastureland and native vegetation ($P>0.05$), and the greatest C-CO_2 emissions were associated with soil from cropland ($P<0.05$). Soil organic carbon and labile carbon concentration, along with soil type, affected soil biological activity. The preliminary results of this study indicated that C-CO_2 emissions from pasture soils can be as low as emissions from native forest soils, while soils under cropping systems have a higher mineralization potential. Our results suggest that organic matter mineralization is a robust indicator for assessing soil health and the nutrient cycling potential across different land uses.

[1]PhD Candidate, São Paulo State University, Jaboticabal, SP, BR (camilaessouza@unesp.br); [2]Professor – Forage Agronomy, University of Florida, North Florida Research and Education Center-Marianna, Marianna, FL (dubeux@ufl.edu); [2]Post-doctoral associate, University of Florida, North Florida Research and Education Center-Marianna, Marianna, FL (ig.limabretas@ufl.edu); Ph.D. Candidate & Graduate Research Assistant, University of Florida, North Florida Research and Education Center-Marianna, Marianna, FL (ldantasqueiroz@ufl.edu); [2]Ph.D. Biological Food/Agronomy Scientist II, University of Florida, North Florida Research and Education Center-Marianna, Marianna, FL (ruizm001@ufl.edu); [2] Research Coordinator Beef and Forage program, University of Florida, North Florida Research and Education Center-Marianna, Marianna, FL (lizagarcia@ufl.edu); [3]Post-doctoral associate, University of Winsconsin, Madison, WI (bizzuticremo@wisc.edu) [4]Professor – Agronomy Department, Federal Rural University of Pernambuco, Recife, PE, BR (mariolirajunior@gmail.com).

THE BIG AND LITTLE OF CORN SILAGE

Courtney A. Harle1, Fernando E. Miguez1, and Shelby M. Gruss[1]

Silage corn (Zea maize L.) production is critical for U.S dairy producers, providing a high-energy, storable feed source rich in digestible fiber and carbohydrates while also offering ease of production. Short stature corn has growing interest across the agricultural industry as a solution to lodging and increased late-season field access, which increases the resiliency and sustainability of the production system. These advantages could also positively impact silage systems. Additionally, short stature corn could increase the leaf-to-stem ratio, potentially increasing nutritive quality, as seen in brachytic dwarf sorghum forages. This study aims to assess the potential of short stature corn hybrids as a sustainable alternative by comparing their performance to traditional (tall) corn hybrids in a silage system. Three traditional and two short stature corn silage hybrids were planted at 34,000 and 40,000 seeds /acre in a randomized complete block design with four replications. Each hybrid was evaluated for biomass production, harvest index, green leaf number, and nutritive value. Preliminary data indicates short stature corn performed equivalent to the traditional corn silage hybrids for biomass productivity across populations. However, initial trends suggest that short stature hybrids may perform better at higher populations, while tall hybrids appear more suited to lower populations. Despite their height, short stature corn can produce as much biomass and as many leaves as traditional (tall) hybrids, showing great promise in a silage system.

[1]Department of Agronomy, Iowa State University, Ames, IA 50011 USA.

PREPARING TOMORROW'S FORAGE INDUSTRY LEADERS: STUDENT MIX & LEARN SUCCESS

K. Klingner[1], K.M. Mason[1], D.H. Tigue[2], S.L. Dillard[3], W.B. Smith[3], and S.M. Justice[4]

Undergraduate and graduate students comprise a significant portion of attendees at the American Forage & Grassland Council's Annual Meeting. While these students can showcase their work through competitions and poster or oral presentations, it is equally vital to provide a dedicated platform for networking and professional development tailored to future leaders in the forage industry. From 2022 to 2024, approximately 90 students participated in the Student Mix & Learn, receiving focused guidance and skill-building for pursuing forage industry jobs. This event features networking sessions with current and former AFGC Board members, professional development activities like resume critiques and interview practice, and a fun game of BINGO with forage-themed prizes. Over three years, participants have comprised 44% undergraduates, 52% graduate students, and 4% others. Post-event evaluations collected 50 responses, showing a 20% increase in students' confidence and knowledge about resume building and interview skills. The event earned an average rating of 4.6 out of 5 for both usefulness/relevance and overall satisfaction. Participants especially appreciated the balance between professional growth and a relaxed, engaging atmosphere. The Student Mix & Learn remains a cornerstone of the annual meeting, equipping students and emerging professionals with the tools to succeed in the forage industry.

[1] Klingner: Graduate Assistant (kklingne@vols.utk.edu), Mason: Assistant Professor/Extension Specialist, Animal Science, University of Tennessee, Knoxville, TN. [2] Tigue - Technical Sales Specialist, Provimi North America; [3] Dillard: Associate Professor/Extension Specialist, Smith: Associate Professor, Animal Sciences, Auburn University, Auburn AL; [4] Justice - Assistant Professor/Extension Specialist, Animal Science, University of Arkansas, Little Rock, AR.

APPROACHES TO REDUCING THE SIZE OF CORED HAY SAMPLES

K.A. Trimble, W. Peck, and C.D. Teutsch[1]

Abstract

A Universal Sample Splitter could offer an efficient and convenient way to accurately subdivide cored hay samples. The objective of this experiment was to evaluate three approaches to reducing sample size. The five treatments were hand grab, Universal Sample Splitter with 1.5", 3.0", and 4.5" openings, and quartering. Following division, samples were oven-dried, ground to pass through a 2- and 1-mm screen using Wiley and Udy sample mills, respectively. The ground tissue was then analyzed for crude protein (CP), acid detergent fiber (ADF), neutral detergent fiber (NDF), and in-vitro dry matter digestibility (IVDMD) using near infrared spectroscopy. Results indicated minimal differences across treatments, with all residual values remaining under 1% for CP, ADF, NDF, and IVTDMD. While crude protein and acid detergent fiber showed no significant differences, quartering yielded slightly higher residuals compared to other methods. As expected, the hand grab method exhibited the quickest processing time, while quartering took more than double the time of the other treatments. Although the findings indicate that both the hand grab method and the Universal Sample Splitter can provide efficient and accurate approaches for reducing the size of cored hay samples, we have concerns about recommending the hand grab method due to the heterogenous nature of the particle size and distribution in cored hay samples.

<u>Introduction:</u> The recommendation for preparing hay samples for nutritive value analysis is to grind the entire sample. Grinding the entire cored hay sample ensures an accurate estimate of the nutritive value. However, grinding larger cored hay samples can be inefficient due to longer grinding times and increased wear on grinders. Subdividing samples could reduce grinding time and wear on equipment. However, due to uneven particle sizes in the cored hay samples (Fig. 1), separating samples accurately can be difficult. The objective of this study was to evaluate the accuracy of a Universal Sample Splitter (riffle splitter) for dividing cored hay samples.

<u>Materials and Methods:</u> An experiment evaluating the accuracy of a Universal Sample Splitter, Model SP-2 (Gilson Company, Inc., Lewis Center, OH) (Fig. 2), for dividing cored hay samples was completed at the University of Kentucky Research and Education Center located in Princeton, KY. The experimental design was a randomized complete block with four replications. The entire experiment was repeated. Eight grass hay bales were sampled with a Star Uni-Forage Sampler with spiral assist (Star Quality Samplers, Inc, Irricana, AB, Canada) with three samples being collected per bale. Samples were oven-dried for three days at 131 degrees Fahrenheit, then, randomly assigned one of the five treatments. Five treatments were hand grab, Universal Sample Splitter (USS) with 1.5", 3.0", and 4.5" openings, and quartering (Undersander et al., 1993). The experiment was conducted on 23-May-2024 and then repeated on 5-June-2024.

---

[1] K.A. Trimble, undergraduate, Murray State university, Murray, KY; W. Peck, undergraduate, University of Kentucky, Lexington, KY; and C.D. Teutsch, Forage Specialist, UK Research and Education Center, Princeton, KY.

All treatments resulted in splitting initial samples into two separate subsamples, labeled A and B. For the hand grab treatment, a initial weight of the initial sample was taken, and then handfuls were pulled until approximately half of the initial weight had been removed. When using the Universal Sample Splitter, the half-inch bars of the unit were flipped to achieve the required spacings (Fig. 2). The manufacturer's recommendation for bar spacing when using the Universal Sample Splitter is to determine the largest particle length in your sample and then multiple by two or three (Gilson Company, Inc., 2022) . For example, if the maximum particle length was 0.5-inches then the optimal bar spacing would be 1 to 1.5 inches. Samples were evenly distributed in the hopper of the Universal Sample Splitter and released using the hopper gate lever (Fig. 3). If bridging occurred, then a soft-bristled 2-inch paint brush was used to gently push the entirety of the sample through the corresponding openings. Quartering was accomplished by evenly spreading the sample into a 20-inch circle with quarter markings that was drawn onto a poster board. The sample was then separated into quarters using a ruler (Fig. 3). Diagonal quarters were combined to create samples A and B. Weights, volumes, and times were recorded for all subsamples.

Subsamples were dried in a forced air oven at 131 degrees Fahrenheit for 72 hours. Samples were then ground to pass through a 2- and 1-mm screen using Wiley (Thomas Scientific, Swedesboro, NJ) and Udy (Udy Corp., Fort Collins, CO) sample mills, respectively. Ground samples were analyzed for crude protein (CP), acid detergent fiber (ADF), neutral detergent fiber (NDF), and in vitro true dry matter digestibility-48 hour (IVTDMD-48) using the grass hay (small ring cup) equation from the NIRS Forage and Feed Consortium (Berea, KY). Guidelines for the optimal use of NIRSC equations were followed (McIntosh, 2022).

Residual values were calculated by subtracting subsample A from subsample B. The absolute value of the residuals were used for the statistical analysis. Data were analyzed across trials using the General Linear Model procedure from SAS (SAS Institute, Cary, NC). Means were separated using Fisher's Protected Least Significant Difference (LSD).

Results and Discussion: No trial-by-treatment interactions occurred. Therefore, results are averaged across the trials. Crude protein percentages were similar for all treatments ($P > 0.98$ (Table 1). Acid detergent fiber was not impacted by sampling method ($P > 0.29$), however, quartering resulted in a slightly higher numerical residual than any other treatment (Table 1). Sampling menthod did not impact NDF ($P > 0.12$). In-vitro dry matter digestibility was impacted by subsampling method ($P = 0.056$). Quartering had the highest residual (1.28%) and different from the riffle splitter with the 3-inch opening (0.47%) (Table 1).

In a commercial lab where high throughput is essential, the time spent subdividing samples would be crucial. Our data indicate that the quartering method took the longest ($P < 0.001$) and was more than more than two-times longer than the other (Table 1). As expected, the hand grab method and the Universal Sample Splitter with a 4.5" opening took the least amount of time ($P < 0.001$).

Summary and Implications: Overall, treatment differences were minimal for the four constituents evaluated; less than 1% differences in the respective residuals. Although our data indicate that the hand grab method was as or more accurate as quartering, we have concerns about recommending this method due to the heterogenous nature of particle size distribution in cored hay samples. In contrast, the Universal Sample Splitter could offer a convenient and accurate approach to subdividing large, cored hay samples. This method also offers a more rapid alternative to quartering. This study was conducted with grass hay and needs to be repeated with legume hay.

Literature Cited:

Gilson Company, Inc. 2022. Operator manual for universal sample splitters. Available online at https://gilsonco.sharepoint.com/:b:/s/ExternalResourceDocuments2/EXOTRRI5NFtCoQA54mwhDTsBE6fkRK4acPEgv2_vJJ8Yhg?e=eOQ94v.

McIntosh, D., B. Anerson-Husmoen, R. Kern-Lunbery, P. Goldblatt, R. Lemus, T. Griggs, L. Bauman, S. Boone, G. Shewmaker, and C. Teutsch. 2022. Guidelines for the Optimal Use of NIRSC Forage and Feed Calibrations in Membership Laboratories (Second Edition). The UT Press, Knoxville, TN.

Undersander, D., D.R. Mertens, and N. Thiex. 1993. Forage Analyses Procedures. NFTA, Stuart, FL. Available online at https://www.foragetesting.org/_files/ugd/24f64f_4502c4143ec34a6ba1c4940c6147fd07.pdf.

Table 1. Impact of sample size reduction treatment on the crude protein, acid detergent fiber, neutral detergent fiber, in vitro dry matter digestility-48 hr residuals, and separation time.

Sample Size Reduction Treatment	CP* Residual	ADF Residual	NDF Residual	IVTDMD-48 hr Residual	Sample Reduction Time
	------------	------------	%	------------	sec
USS†-1.5 in	0.25a‡	0.56a	0.78a	0.81ab	95
USS-3.0 in	0.24a	0.30a	1.0a	0.47b	99
USS-4.5 in	0.28a	0.46a	0.54a	0.74ab	71
Hand grab	0.22a	0.38a	0.35a	0.58b	57
Quartering	0.27a	0.68a	0.71a	1.28a	170
Prob > F	0.99	0.30	0.13	0.06	< 0.001
LSD (0.05)§	ns	ns	ns	0.56	24

*CP, crude protein; ADF, acid detergent fiber; NDF, neutral detergent fiber; IVTDMD-48 hr, in vitro true dry matter digestibility-48 hr.

†USS, universal sample splitter.

‡Means within a column followed by the same letter are not different according to Fisher's Protected Least Significant Difference, alpha = 0.05.

§LSD (0.05), Fisher's Protected Least Significant Difference, alpha = 0.05.

Figure 1. In cored hay sample there are a range of particle sizes. Smaller particles tend to migrate to the bottom of the bag while larger particles remain higher. This range in particle size and migration make accurately subdividing cored hay samples difficult.

Figure 2. Gilson Universal Sample Splitter, Model SP-2 (Gilson Company, Inc., Lewis Center, OH) (left). Opening size is adjusted by flipping ½ inch bars back and forth until the desired spacing is achieved (right).

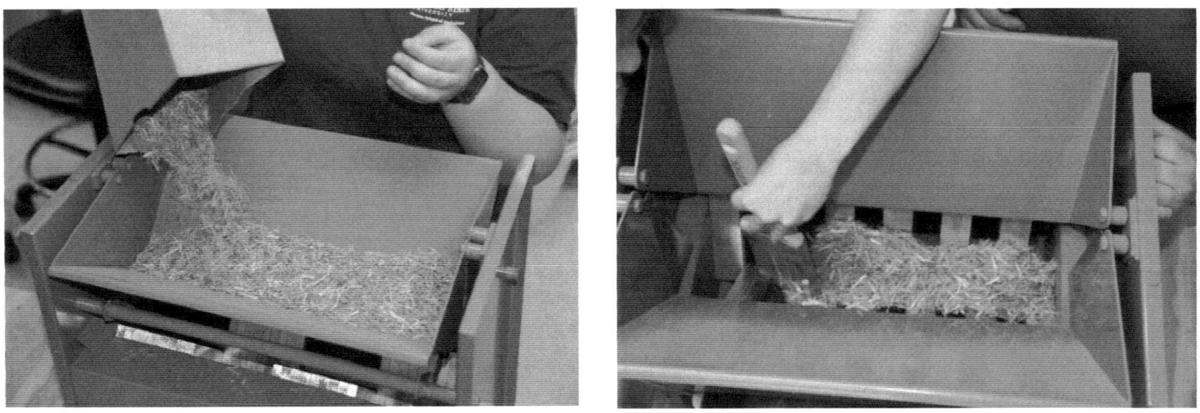

Figure 3. Samples were evenly distributed in the hopper box before being released using the hopper release lever (left). If the samples bridged, a soft bristled brush was used to sweep the sample through the appropriate openings.

USDA-ARS FORAGE-ANIMAL PRODUCTION RESEARCH UNIT: A NEW LABORATORY TO SERVE THE AMERICAN FORAGE INDUSTRY

Michael Flythe[1]

Design is complete on a new laboratory dedicated to the study of forages and the animals that utilize them. The federally funded project is a partnership between the USDA, Agricultural Research Service (ARS) and the University of Kentucky's Martin-Gatton College for Agriculture, Food, and the Environment (MGCAFE). The location will be on the University of Kentucky campus in Lexington, just south of current MGCAFE research and educational facilities. The two story, 39,000 gsf structure will include 13 principal investigator labs dedicated to the study of animals, plants, microorganisms, and soils within the forage-animal research area. There will be an analytical chemistry core and a headhouse, and greenhouses for the growth of forage plants and industrial hemp, which is a new part of the ARS mission. The labs will be staffed by a consolidated team of ARS and MGCAFE researchers, a partnership that already has 20 years of success. The intention for this new facility is that we may continue to serve the cattle, small ruminant, and equine forage industries for many years to come.

[1]Research Leader, Forage-Animal Production Research Unit.

EVALUATING NITROGEN RATE AND TIMING EFFECTS ON TALL FESCUE YIELD AND NUTRITIVE VALUE

B.L. Hendrix and C.D. Teutsch[1]

Abstract

Application of late-season nitrogen (N) is a routine practice in turfgrass production, causing an extended green period in the fall without excessive shoot growth. This results in higher levels of carbohydrate reserves and enhanced root growth compared with spring-summer N fertilization. In forage production however, there is limited research showing the impact of fall N application on grass growth in hayfields and pastures. The objective of this study was to determine the impact of fall and spring N applications on the dry matter yield and nutritive value of tall fescue. This experiment was conducted at the University of Kentucky Research and Education Center (UKREC) in Princeton, KY in 2022-23, and repeated in 2023-24. The experimental design was a complete factorial arranged in four randomized complete blocks. Plots received N either in the fall, in the spring, or in both fall and spring. Fall rates were 0, 40 or 80 lb N/A, and spring rates were 0, 40, 80 or 120 lb N/A, all as urea (46-0-0). Plots were harvested in May (2023 and 2024) and again in July (2023) or August (2024) using a forage harvester equipped with load cells. Subsamples were collected for dry matter and nutritive value analysis from all harvests. The data indicate that fall N increased first harvest yield but had a lesser impact on second harvest yield for both experiments. Total yield exhibited a linear response up to 200 lb total N/A in 2023. The spring N rate impacted CP, ADF, and NDF values, but these same constituents were less affected by fall N rate. Plots receiving fall N also initiated growth earlier in the spring. Optimum N application timing may depend on whether the forage is being hayed or grazed.

Introduction: Tall fescue (*Schedonorus arundinaceus* (Schreb.) Dumort., nom. cons.) is the most important cool season grass species in the United States (Hoveland, 2009), providing pasture for more than 10 million brood beef cows (USDA-NASS). This grass is known to be long-lived and robust, largely due to its mutualistic relationship with an endophytic fungus (Hoveland, 2009). Late-season nitrogen (N) fertilization is a practice that has long been implemented in turfgrass production systems (Street et al., 2015). This practice involves applying N between September and December, and results in an extended green period in the fall without stimulating excessive shoot growth (Street et al., 2015). This longer green period results in higher levels of carbohydrate reserves and enhanced root growth when compared with traditional spring-summer N fertilization programs. In addition to these benefits, sods fertilized in late fall have been shown to green up two to six weeks earlier in the spring (Street et al., 2015). However, limited research has considered the effect of N application timing on grass growth in hayfields and pasture systems. The objective of this study was to determine the impact of fall and spring N applications on the dry matter yield and nutritive value of tall fescue.

[1]Senior Laboratory Technician, UKREC, Princeton, KY (brittanyhendrix@uky.edu); Associate Professor for Forages Extension and Research, UKREC, Princeton, KY (chris.teutsch@uky.edu).

Materials and Methods: An experiment evaluating the impact of N rate and application timing on the dry matter production and nutritive value of tall fescue hay was begun at the University of Kentucky Research and Education Center located in Princeton, KY in the fall of 2022 and continued through the spring and summer of 2023. This study was then repeated in the fall of 2023 through the spring and summer of 2024. The experiment was laid out in four randomized complete blocks. Plots were 10 x 20 ft, with 5 ft alleys, and were placed within an existing tall fescue hayfield. Treatments were conducted in separate locations for the 22-23 and 23-24 experiments. These treatments consisted of a complete factorial of three fall N rates (0, 40 or 80 lb N/A) and 4 spring N rates (0, 40, 80 or 120 lb N/A), resulting in plots receiving either N in the fall, in the spring, or both fall and spring. Urea (46-0-0) was hand-broadcast over the surface of each plot on 2 December 2022 and 4 April 2023, then on 5 December 2023 and 20 March 2024 for the first and second experiments, respectively.

Beginning in March (2024) or April (2023) after the spring N application, compressed sward heights were collected using a rising plate meter. Three measurements were taken down the center of each plot and the values recorded. The differences between plots were analyzed as a means to compare density. This process was continued weekly until the second harvest in July (2023) or August (2024). A 5 ft wide strip was harvested down the center of each plot using a self-propelled sickle bar forage harvester equipped with load cells. Yield values from each plot were recorded and later used to calculate estimated forage yield per acre. Plots were harvested on 3 May 2023 and 13 July 2023 for the first experiment, and on 10 May 2024 and 27 August 2024 for the second experiment. Subsamples were collected from each plot for dry matter and nutritive value determinations at each harvest. These subsamples were dried at 131 F for 72 hours before being ground through a Wiley mill at 2 mm, followed by a UDY cyclone mill at 1 mm. Nutritive value was determined using near infrared spectroscopy (McIntosh, 2022). Data were analyzed using the polynomial regression (SigmaPlot 15.0) and General Linear Model (GLM) procedures (SAS, Cary, NC).

Results and Discussion: Year by treatment interactions occurred so data are being presented by year. No significant fall N by spring N interactions were found for either experiment, so the main effects are being presented. In 2023, first harvest dry matter (DM) yield response to N rate was quadratic for both fall and spring N applications, while both second harvest DM responses were linear (Figs. 1 and 3). In 2024, both fall and spring applications showed a linear DM response for the first and second harvests (Figs. 2 and 4). Total DM yield exhibited a linear response in 2023 and a weak quadratic response in 2024 up to 200 lb total N/A (data not shown). Plots receiving fall N also initiated growth earlier than those that did not (data not shown). Spring applied N rate impacted CP, ADF, and NDF values (Table 2). These same constituents were less impacted by fall N application (Table 1; $P > 0.05$).

The data indicate that fall N increased first harvest yield but had a lesser impact on second harvest yield for both experiments. Optimum N application timing may depend on whether the forage is grown for hay or pasture. In grazing systems, the earlier spring green up and growth seen in the fall N fertilized plots may be more beneficial as this could increase the number of grazing days for that operation. Since 60% of a livestock producer's total cost is in animal feed (FINBIN, 2023), every additional grazing day is money saved. Moreover, applying

N in the fall may also reduce damage to a producer's fields by eliminating the need to drive over them under wet early spring conditions. For hay producers, a fall N application may increase first harvest yield, while a subsequent spring application may increase both the yield and quality of their second cutting.

Literature Cited:

FINBIN. 2023. Center for Farm Financial Management: University of Minnesota. Report ID: 840858. Retrieved from http://finbin.umn.edu (originally created September 20, 2023).

Hoveland, C. S. 2009. Origin and history. p. 3-10. *In* H.A. Fribourg, et al. (eds) Tall Fescue for the Twenty-first Century. Agron. Monogr. No. 53. ASA-CSSA-SSSA. Madison, WI.

McIntosh, D., B. J. Anderson-Husmoen, R. Kern-Lunbery, P. Goldblatt, R. Lemus, T. Griggs, L. Bauman, S. Boone, G. Shewmaker, and C. Teutsch. 2022. Guidelines for Optimal Use of NIRSC Forage and Feed Calibrations in Membership Laboratories (Second Edition). The University of Tennessee Press, Knoxville, TN.

Street, J.R., P.J. Sherratt, and K.T. Danneberger. 2015. The benefits of late season fertilization. OSU Spots Turf Management. Available at https://u.osu.edu/athleticfieldmanagement/2015/11/24/the-benefits-of-late-season-fertilizer/

U.S. Department of Agriculture. *Quick Stats.* Washington, DC: U.S. Department of Agriculture, National Agricultural Statistics Service. Available online at https://quickstats.nass.usda.gov/

Table 1. Crude protein (CP), acid detergent fiber (ADF) and neutral detergent fiber (NDF) values for all harvests as related to fall N rate, averaged across all spring N rates. The least significant difference (LSD) value is shown unless that value was not significant (ns).

			Fall Nitrogen Rate (lb N/A)			
			0	**40**	**80**	**LSD (0.05)**
			%			
2022-23	Harv 1	**CP**	13.55	13.47	13.68	ns
		ADF	31.78	32.20	31.74	ns
		NDF	57.64	58.68	57.62	ns
	Harv 2	**CP**	8.80	8.84	9.05	ns
		ADF	37.04	36.95	37.13	ns
		NDF	62.55	62.47	62.30	ns
2023-24	Harv 1	**CP**	12.33	12.27	13.45	ns
		ADF	37.05	37.26	37.00	ns
		NDF	63.65	64.19	63.57	ns
	Harv 2	**CP**	8.40	8.76	9.16	0.32
		ADF	39.43	39.60	40.03	ns
		NDF	64.99	65.15	66.15	0.84

Table 2. Crude protein (CP), acid detergent fiber (ADF) and neutral detergent fiber (NDF) values for all harvests as related to spring N rate, averaged across all fall N rates. The least significant difference (LSD) value is shown unless that value was not significant (ns).

			Spring Nitrogen Rate (lb N/A)				
			0	**40**	**80**	**120**	**LSD (0.05)**
			%				
2022-23	Harv 1	**CP**	10.49	13.01	14.61	16.16	0.58
		ADF	33.14	32.05	31.64	30.81	0.78
		NDF	58.95	58.16	57.96	56.86	1.30
	Harv 2	**CP**	9.06	8.52	8.80	9.21	ns
		ADF	38.23	37.21	36.53	36.19	0.66
		NDF	63.64	62.30	62.08	61.75	1.22
2023-24	Harv 1	**CP**	11.09	11.41	13.43	14.80	1.64
		ADF	38.05	37.71	36.48	36.17	0.87
		NDF	65.07	64.55	62.96	62.63	1.28
	Harv 2	**CP**	8.26	8.42	9.10	9.30	0.36
		ADF	39.36	39.73	39.79	39.85	ns
		NDF	64.73	65.12	66.01	65.85	0.97

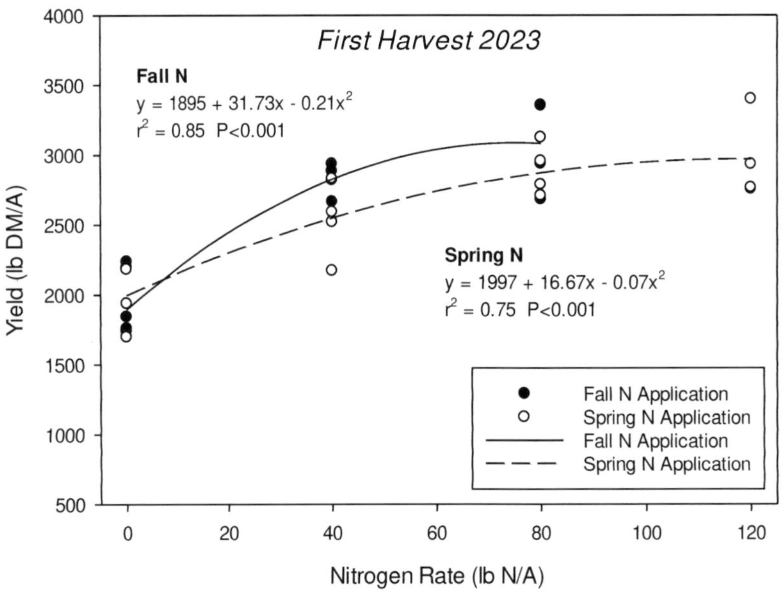

Figure 1. Impact of fall and spring N rate (lb N/A) on first harvest yield in 2023 (lb DM/A).

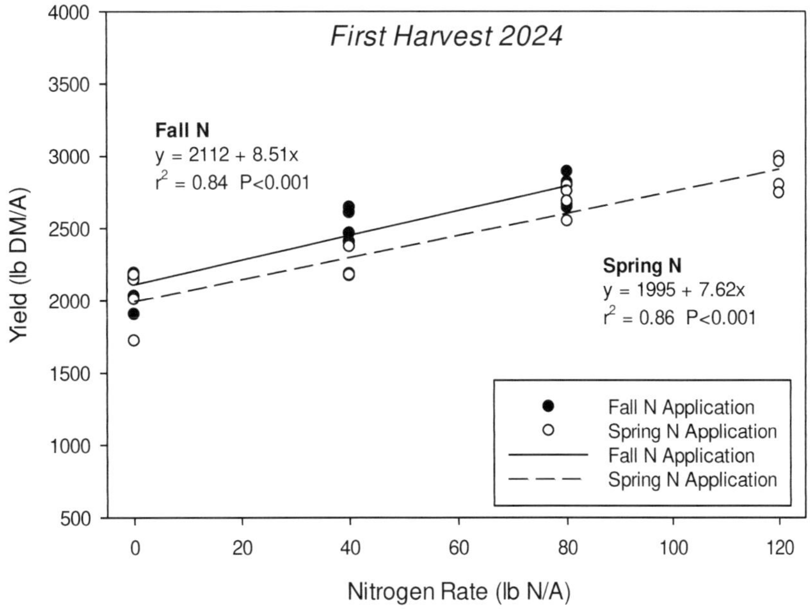

Figure 2. Impact of fall and spring N rate (lb N/A) on first harvest yield in 2024 (lb DM/A).

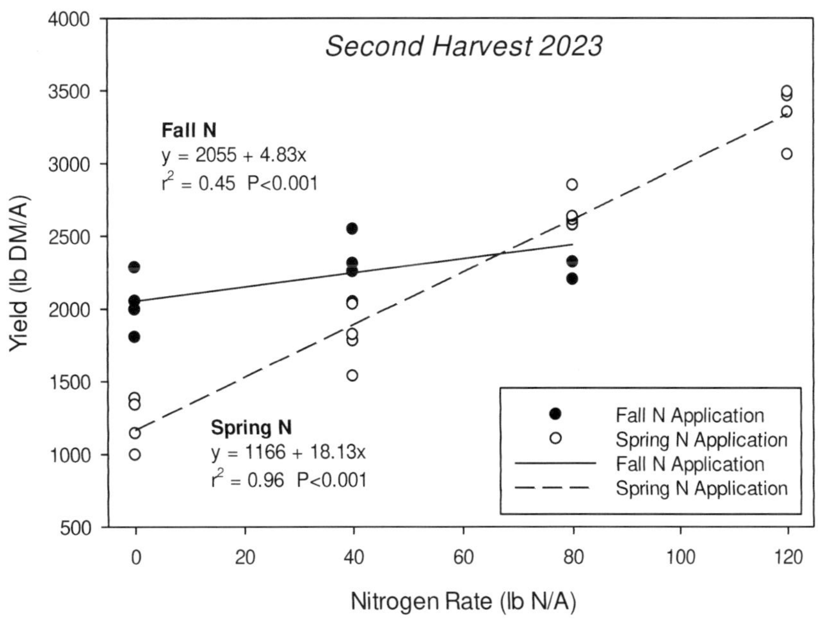

Figure 3. Impact of fall and spring N rate (lb N/A) on second harvest yield in 2023 (lb DM/A).

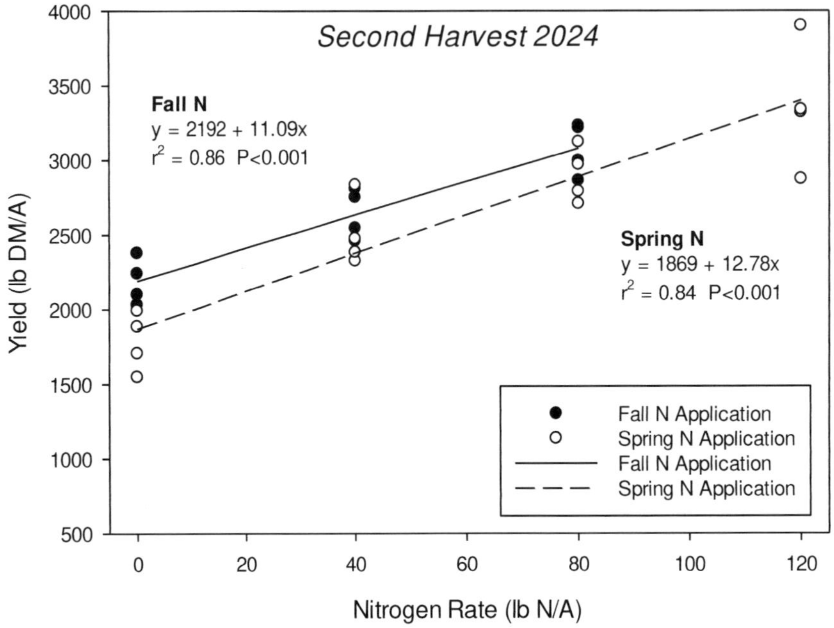

Figure 4. Impact of fall and spring N rate (lb N/A) on second harvest yield in 2024 (lb DM/A).

BALE GRAZING AS TOOL FOR BUILDING SOIL FERTILITY IN LOW INPUT GRAZING SYSTEMS

C.D. Teutsch, W.E. Peck, B.L. Hendrix, and K.R. VanValin[1]

In the southeastern United States, pasture productivity is commonly limited by low soil fertility. High fertilizer prices often limit applications in low input pastoral systems. An alternative to fertilizer may be using bale grazing to distribute fertilizer nutrients contained in hay. In this system bales are uniformly distributed throughout the pasture and allocated to cattle using temporary electric fencing. Little quantitative data are available for the bale grazing in the southern U.S. The objective of this project was to document the impact of bale grazing on changes in soil fertility. This project was conducted at the University of Kentucky Research and Education Center in Princeton, KY. Forty bales with an average weight of 822 lb/bale were uniformly distributed in a 10 A pasture. The pasture was stocked with 61 cow-calf pairs on 07-Mar-23 and animals were removed on 20-Mar-23. Hay rings were used to minimize feeding losses, and temporary electric fencing was utilized to allocate enough hay for a 2 to 3-day period. Soil nutrient levels were documented at the start of the study and one year later by sampling every 20 ft along two permanent transects across the pasture. One season of bale grazing increased soil test P, K, Ca, Mg, and Zn ($P < 0.003$). Soil test P increased from 18 to 22 lb/A ($P < 0.001$) while K increased from 125 to 161 lb/A ($P < 0.001$). Average manure pile density was 1,513 piles/A with a dry weight of 0.70 lb/pile. The initial results of this study indicate that bale grazing may have the potential to change soil test nutrient levels more rapidly than expected.

[1]Associate Extension Professor, UKREC, Princeton, KY (Chris.Teutsch@uky.edu); Undergraduate Intern, UKREC, Princeton, KY (Whitney.Peck@uky.edu); Laboratory Technician Senior, UKREC, Princeton, KY (brittanyhendrix@uky.edu); Assistant Extension Professor and Beef Specialist, UKREC, Princeton, KY (Katie.VanValin@uky.edu).

IMPACT OF *OPUNTIA FICUS-INDICA* CHRONOSEQUENCE ON SOIL CARBON MINERALIZATION AND FERTILITY

Giuseppe Greco[1], Giorgia Liguori[1], Francesco Gargano[1], Paolo Inglese[1], Camilla Souza de Souza[2], Martin Ruiz Moreno[2], and José C.B. Dubeux, Jr[2]

Opuntia ficus-indica, known as cactus pear, is exceptionally well-suited for arid climates, thriving through its Crassulacean Acid Metabolism (CAM) which enhances water use efficiency and enables photosynthesis under drought conditions. This adaptive feature supports both fruit production and use as forage in drought-prone regions such as Africa and South America. This study investigates the impact of orchard age on soil organic carbon (SOC) mineralization across orchards aged 1 to 30 years in Roccapalumba and San Cono, Sicily (Italy). Soil samples from 0- to 15-, 15- to 30-, and 30- to 60-cm layers were incubated at 25°C to measure CO_2 emission as a proxy of microbial activity and the breakdown of organic matter. The results demonstrated that SOC mineralization rates, expressed in mg CO_2-C kg^{-1} soil d^{-1}, tended to increase with the age of the orchard. This enhanced mineralization process is the result of soil C accumulation over time, and it is critical for sustaining soil fertility and promoting sustainable agricultural practices in ecosystems where CAM plants are prevalent. The study highlights the importance of understanding SOC dynamics to optimize soil management practices, aiming to maintain productivity and ecological balance in arid agricultural systems. These insights are crucial for developing targeted strategies to enhance carbon cycling and soil health, providing a foundation for sustainable land use in regions characterized by water scarcity.

[1]Department of Agricultural, Food and Forest Sciences, University of Palermo, Viale delle Scienze, Bldg. 4, 90128 Palermo, Italy; [2]North Florida Research and Education Center, University of Florida. 3925 Highway 71, Marianna, Florida, USA 32446-8091.

EFFECTS OF BIOCHAR ON FORAGE PRODUCTION AND QUALITY

G. Kilpatrick, C. Foote, S. Hayes, and K.M. Dick[1]

Biochar, as a soil amendment, has the potential to enhance agriculture production. However, most biochar research focuses on row crop production as opposed to forage production and quality. Therefore, a small plot experiment was established in a Bermuda grass (*Cynodon dactylon*) and bahiagrass (*Paspalum notatum*) hayfield in which five treatments were applied in a 3x3 meter plot with five replicates each for a total of 25 plots. Treatments were arranged in a Latin Square design. Treatments included a Control with no amendments at all, a Biochar Only (BCO) treatment, a Low Biochar (LB) treatment with 75% chicken litter and 25% biochar, a High Biochar (HB) treatment with 50% chicken litter and 50% biochar, and a Litter Only (LO) treatment. Amount of treatment applied was equivalent to 2 tons/ac. Dry forage biomass was collected and measured for the second and third hay cuttings in summer 2024. Forage quality was analyzed for the second hay cutting. There was no statistical difference among treatments for any measured variable. However, on average, the BCO treatment resulted in 10% less dry forage biomass compared to the Ctrl while the LB treatment had 7% more. When mixed with chicken litter, biochar had no effect on forage production or quality though applications of biochar alone could have a negative impact on forage production in the first year of application.

[1]Undergraduate Student, University of Arkansas at Monticello, Monticello, AR (gk049042@uamont.edu); Undergraduate Student, University of Arkansas at Monticello, Monticello, AR (cf045508@uamont.edu); Drew County Extension Agent, University of Arkansas Division of Agriculture, Monticello, AR (shayes@uada.edu); Assistant Professor, University of Arkansas at Monticello, Monticello, AR (bridgesk@uamont.edu).

EVALUATION OF FALL AND SPRING NITROGEN RATES ON CEREAL RYE FORAGE CRUDE PROTEIN USING NDVI AND CANOPEO TO MAKE INFIELD NITROGEN RATE DECISIONS

J.M Hartschuh[1], K.M. Stahl[2], and A.M Gahler[3]

Nitrogen management is critical to maximizing winter annual cereal rye yield and quality. This study was conducted to determine if better spring nitrogen decisions can be made by using remote sensing technology to determine spring nitrogen stats. Fall planted cereal rye received fall and spring nitrogen applied in a randomized complete block split-plot design. Fall nitrogen rates of 0, 30, 60, and 90 lbs/ac were applied as urea. In the fall and spring NDVI readings were taken using a hand-held Greenseeker and ground cover percentages using the Canopeo App. Spring NDVI readings were significantly different for each fall nitrogen rate. When zero lbs/ac of nitrogen was applied the average NDVI reading was 0.40 and when 90 lbs/ac of nitrogen was applied the average NDVI reading was 0.65. Canopeo readings were also significantly different as nitrogen rates increased the Canopeo percentages increased. Spring nitrogen was applied over all fall nitrogen rates at the rates of 25, 50, 75, and 100 lbs/ac as urea. Yield was maximized when a minimum of 30 lbs/ac of fall nitrogen was applied. Crude protein increased for each spring nitrogen rate applied. When higher fall nitrogen rates are applied, lower spring nitrogen rates are needed to maximize yield and quality. 90 lbs/ac of fall nitrogen and 25 lbs/ac of spring nitrogen maximized yield while 90 lbs/ac of fall nitrogen and 50 lbs/ac of spring nitrogen were needed to maximize crude protein. Sensor technology can be used to assist in spring nitrogen rate decisions.

[1]Field Specialist, Dairy Management and Precision Livestock, The Ohio State Univ., Fremont, OH (Hartschuh.11@osu.edu); [2]Extension Educator Agriculture and Natural Resources, Crawford County, The Ohio State University, Bucyrus, OH (stahl.221@osu.edu); [3]Extension Educator Agriculture and Natural Resources, Sandusky County, Fremont, OH (Gahler.2@osu.edu).

EVALUATION OF NITROGEN RATES FOR IRRIGATED BERMUDAGRASS HAY PRODUCTION IN ARKANSAS

L. Blythe and B. Finch[1]

Hay production is a vital for providing ample high-quality supplemental forage for grazing animals during months when pasture forage is limited. As efficient irrigation technologies become more available and affordable there is an increase in irrigated forage production. While most management strategies are parallel to non-irrigated production, the potential yields and efficiency increase. Nutrient demands, specifically nitrogen (N), increase as yield potential increases, however, most forage N recommendations have been developed based on non-irrigated production systems and may not maximize yield potential in irrigated production. This study evaluated the influence of N application rate on bermudagrass dry matter yield and quality under irrigated management. Eight treatments were established in Midland bermudagrass with 5 N rates (50 – 250 lb N/acre) applied at green-up and following each cutting, an un-fertilized check, and two single application rates to monitor for residual nitrogen influence. The average harvest cycle length was 32 days from fertilization to harvest, with no irrigation needed in the first and second harvest cycles due to precipitation amounts. Nitrogen application increased yield production across the season except in the first harvest, with an average increase of 3.3 Tons DM/acre for the season. On average crude protein and relative feed quality increased with N rate up to 200 lb N/acre. Results of the first year of the study provide insight into the potential need for increasing N rate recommendations for hay production under irrigation.

[1]Soil Fertility Specialist, University of Arkansas System Division of Agriculture, Cooperative Extension Service, Little Rock, AR (bfinch@uada.edu); Greene Co. Agent, University of Arkansas System Division of Agriculture, Cooperative Extension Service, Paragould, AR (lblythe@uada.edu);

DEMONSTRATING NUTRIENT NEEDS FOR LATE WINTER PLANTED SPRING OATS

S. Stone[1], B. Finch[2], K. Simon[3], and D. Kennedy[4]

Planting summer annuals, such as spring oats in late winter, can provide a valuable forage resource to fill gaps in the grazing period before warm-season forages are available. However, achieving maximum forage biomass in such a condensed growing season requires careful nutrient management. Nutrient-limiting conditions can significantly hinder the productivity of forage crops, reducing the ability to produce sufficient biomass for grazing. This demonstration aimed to compare the response of late-winter planted spring oats to nitrogen (N), phosphorus (P), Potassium (K), and Sulfur (S) individually and in combinations. This study provides valuable insights for forage managers seeking to optimize nutrient applications in short growing seasons, ensuring robust forage production to support livestock grazing during critical seasonal gaps. Results indicated that N was the most influential nutrient, significantly increasing forage biomass production by an average of 1,995 lb. dry matter per acre, with the greatest response observed when N was applied in combination with K, yielding a 2,383 lb increase in dry matter. Phosphorus application, on the other hand, generally reduced yields, with the exception of when it was applied alongside K, where a 545 lb increase in dry matter was observed. The observed decrease in yield from N and P combination may be attributed to temporary changes in local soil pH and increased volatilization of urea.

[1]St. Francis County Agent, University of Arkansas System Division of Agriculture, Cooperative Extension Service, Forrest City, AR (sastone@uada.edu); [2]Soil Fertility Specialist, University of Arkansas System Division of Agriculture, Cooperative Extension Service, Little Rock, AR (bfinch@uada.edu); [3]Animal Science Instructor, University of Arkansas System Division of Agriculture, Cooperative Extension Service, Little Rock, AR (ksimon@uada.edu); [4]Associate Dean and Professor of Animal Science, Arkansas State University, Jonesboro, AR (dkennedy@astate.edu).

EVALUATING THE NUTRITIONAL STATUS OF ALFALFA STANDS IN THE MID-SOUTH

W.R. Fleming, C.D. Teutsch, E.L. Ritchey, and J.H. Grove[1]

Alfalfa (*Medicago sativa* L.) is a forage legume highly regarded worldwide for its high nutritional value when fed to livestock. National alfalfa yields followed a similar pattern to other crops, such as corn and soybean, increasing exponentially beginning in the 1950s. However, in the 1980s alfalfa yields began to plateau at 3.3 tons/acre, where they remain today. The objective of this study was to determine soil fertility's role in the observed yield plateau, specifically in the mid-south region of the United States. This survey is a subset of a larger survey funded by the USDA. A total of 114 alfalfa stands were sampled in 2022 and 2023. Soil samples to a depth of 4 and 6 inches, tissue samples, and a relative yield measurement were obtained from each stand. Approximately 55% of stands were outside the ideal pH range. According to the University of Kentucky's soil test recommendations for alfalfa, the majority of stands were high in soil test phosphorus (P) regardless the soil test depth. However, the 4 inch sampling, soil test potassium (K) was high in 45% of stands, moderate in 30%, low in 23% and very low in 2%. Soil test K status shifted to 27% high, 33% moderate, 38% low, and 2% very low when sampled to 6 inches. Tissue analysis indicated that tissue magnesium (Mg), sulfur (S) and K were below sufficiency in 35%, 25% and 23% of sampled stands, respectively. However, our data indicated that soil fertility was not the driver of the observed yield plateau in alfalfa, though some stands may benefit from improved nutrient management.

[1]Graduate Student, Department of Crop and Soil Sciences, University of Georgia, Tifton, GA; Associate Professor, Department of Plant and Soil Sciences, University of Kentucky, Princeton, KY; Associate Professor, Department of Plant and Soil Sciences, University of Kentucky, Princeton, KY; Professor, Department of Plant and Soil Sciences, University of Kentucky, Princeton, KY.

THE EFFICACY OF POLY4 AS A K AND S SOURCE FOR ALFALFA

W.R. Fleming, C.D. Teutsch, E.L. Ritchey, and J.H. Grove[1]

Polyhalite is a mineral consisting of potassium (K), calcium (Ca), and magnesium (Mg) as sulfate salts. Recently, the discovery of large deposits in the United Kingdom has created interest in its possibility as a fertilizer. POLY4 is fertilizer derived from polyhalite containing 14% K_2O, 12% Ca, 3.6% Mg and 19% S. The objective of this study was to determine if POLY4 is comparable to other common K and S sources used in alfalfa production. A secondary objective was to determine if there were any slow-release properties to POLY4 as regards these nutrients. Trials were conducted in 2023 at the University of Kentucky's research farms in Lexington and Princeton, KY. Treatments consisted of an untreated control (UTC), muriate of potash (MOP), MOP/gypsum, 10/90 ratio of POLY4/MOP, 20/80 POLY4/MOP, and 30/70 POLY4/MOP applied at a rate of 200 lb K_2O/acre. The Lexington trial provided 5 harvests while Princeton trial provided only 4. Only the Lexington 5th harvest showed a difference in yield due to fertilizer source, where the 30/70 POLY4/MOP gave a greater yield than the UTC. The definitive cause of this yield response was undetermined, but potentially could be the result of gradual S release by the POLY4. Release of K by POLY4 appears to be immediate, rather than over time. These results indicate that POLY4 was a K and S source comparable to other fertilizers commonly used in alfalfa production.

[1]Graduate Student, Department of Crop and Soil Sciences, University of Georgia, Tifton, GA; Associate Professor, Department of Plant and Soil Sciences, University of Kentucky, Princeton, KY; Associate Professor, Department of Plant and Soil Sciences, University of Kentucky, Princeton, KY; Professor, Department of Plant and Soil Sciences, University of Kentucky, Princeton, KY.

EFFECTIVENESS OF PLANT IDENTIFICATION APPS IN IDENTIFYING COMMON FORAGE SPECIES

A.M. Grev and R.N. Herron[1]

A wide variety of mobile phone applications offer users the potential to quickly and easily identify plant species; however, the accuracy of mobile phone applications at identifying plant species under field conditions in hay and pasture systems is unclear. The objective of this project was to test the accuracy of nine popular phone identification applications at identifying common plant species found in forage systems. Target species for this project included common forage and weed species located in pastures and hayfields. A total of 30 different plant species (27 broadleaf, 3 grass) were included; all plants were photographed under normal field conditions without removal or disturbance. For each plant species, three unique images were selected, with priority given to images depicting whole plants in a vegetative state. Each image was then individually run through nine different automated identification applications, including: PictureThis, iNaturalist, Seek by iNaturalist, PlantSnap, LeafSnap, PlantNet, Plantum, Google Lens, and Apple Visual Look Up. Identification performance for each image was scored as follows: 4 = top suggestion correct; 3 = second suggestion correct; 2 = third suggestion correct; 1 = genus correctly identified but not species; and 0 = correct identification not provided. Across all applications, 61% of images were identified correctly as the first suggestion and 74% were identified correctly within the first three suggestions. PictureThis was the best-performing application, identifying 94% of images correctly on the first suggestion, followed by Plantum (89%) and iNaturalist (79%). It appears that phone-based plant identification applications can be a useful tool for those wanting rapid identification of plant species in hay and pasture systems.

[1]University of Maryland Extension, WMREC, Keedysville, MD (agrev@umd.edu).

SHARING SCIENCE: DEVELOPING A STRATEGIC SOCIAL MEIDA PLAN FOR FORAGE EDUCATION AND OUTREACH

C. G. Roper[1], Chris Teutsch[2], and Ray Smith[3]

With more than 188 million daily active users in the United States, Facebook has become a critical player in disseminating information to the public and a part of Extension programming. Scientific-based information can be challenging to communicate given the amount of competition for attention on the platform, volatility of social media trends and the ever-changing algorithms deciding what content to highlight. However, developing a strategic social media plan to cut through the clutter and target grazers is a way to put forage education and outreach in the forefront. When developing a strategic social media plan for forage education, remember E-I-E: Engage, Invite and Educate. Content posted on social media should serve the function of either engaging your stakeholders by communicating to them about your activities or research, inviting your stakeholders to future events, or educating your stakeholders about timely forage-related topics. By focusing on content serving one of three functions, you can cut through the social media clutter and provide timely, educational, targeted information to stakeholders and successfully grow your forage education and outreach program.

[1]Master Grazer Coordinator, University of Kentucky, Princeton, KY (caroline.roper@uky.edu); [2]Extension Associate Professor, University of Kentucky, Princeton, KY (chris.teutsch@uky.edu); [3]Extension Professor, University of Kentucky, Lexington, KY (raysmith1@uky.edu).

LEVERAGING RENEWABLE NITROGEN SOURCES FOR SUSTAINABLE FORAGE SYSTEMS

B.C. Pedreira, C.E. Lima, F.F. Nassar; D.W. McIntosh, R.A. Reis, and R.N. Oakes[1]

Forage is the main feed for livestock in Tennessee, and improving pasture productivity is important. Crabgrass (*Digitaria* spp.) is a high-quality annual forage that vigorously grows during the summer and can self-reseed; N fertilization is needed to reach high production and quality. With the demand for N-dependent intensification, plant growth-promoting bacteria has become a promising technology to allow partial replacement of N fertilizers, improve plant nutrition and soil conservation, and promote C sequestration. This project looked at the effects of *Azospirillum brasilense* bacteria (strains Ab-V5 and Ab-V6) on crabgrass production. In the greenhouse, crabgrass seeds (cv. Mojo) were treated with liquid or powder forms of the bacteria and planted in trays. In the field, we tested the same bacteria treatments on crabgrass plots, with or without nitrogen fertilizer (50 or 67 kg/ha). In the greenhouse, plants treated with the liquid bacteria showed higher shoot weights than untreated plants (1.41 g vs. 1.19 ounces), but no differences were seen among the bacterial treatments. Root growth was not affected. In the field, nitrogen fertilizer improved forage protein levels, regardless of bacterial treatment. Yield and forage quality were similar whether nitrogen came from fertilizer or was combined with the bacteria. Using these bacteria shows promise, but more studies are needed to confirm how and when farmers should use them. For now, nitrogen fertilizer remains a reliable way to increase forage yield and quality.

[1]Associate Professor, Univ. of Tennessee, Knoxville, TN (pedreira@utk.edu); Undergraduate Student, State Univ. of São Paulo, Jaboticabal, SP, Brazil; Graduate Student Univ. of Tennessee, Knoxville, TN; Coordinator, Univ. of Tennessee, Knoxville, TN; Professor, State Univ. of São Paulo, Jaboticabal, SP, Brazil; Professor, Univ. of Tennessee, Knoxville, TN.

INVESTIGATING SOIL MICROBES MEDIATING NUTRIENT TRANSFER IN GRASS-LEGUME MIXTURES

A.S. Erhunmwunse[1] and H.L. Liao[2]

Bahiagrass (*Paspalum notatum*) is a widely grown warm-season grass in the Southeastern U.S., but it requires adequate nitrogen fertilization to maintain productivity. Rhizoma peanut (RP) (*Arachis glabrata*), a perennial legume, can complement bahiagrass by improving growth with less N input and enhancing soil microbial activity. We investigated the role of RP in shaping soil microbial communities and nutrient cycling in Florida bahiagrass pastures. Over four years, we conducted field experiments in short-term (<2 years) and long-term (>8 years) bahiagrass-RP systems, alongside on-farm trials across North, Central, and South Florida. A greenhouse study was also carried out to determine how RP influences forage growth, plant phosphorus (P) uptake, and arbuscular mycorrhizal fungi (AMF) activity under low soil P levels. Our findings showed that bahiagrass mixed with RP increased soil microbial diversity and promoted microbes involved in nutrient cycling and organic material break down. Soil fungal communities responded quickly to RP, while bacterial changes became evident with long-term RP integration. On-farm trials revealed that the impact of RP varied by location and season. Our greenhouse study showed the capacity of RP to increase soil P availability and AMF populations in soil and roots. Integrating RP into bahiagrass pastures can support grass pastures by promoting beneficial soil microbes, though its effects may vary based on the system's age, location, and seasonal conditions.

[1]University of Florida, Southwest Florida Research and Education Center, Immokalee, FL;
[2]University of Florida, North Florida Research and Education Center, Quincy, FL.

DEVELOPING AND TESTING 2,4-D RESISTANT RED CLOVER

S.R. Smith[1], G. Olson, S. Carter, L. Araújo, T. Lewis, M. Barrett, and G. Roberts

Dr. Norm Taylor initiated a 2,4-D tolerant red clover breeding program over 30 years ago by crossing a known tolerant line from Dr. Ken Quesenberry (UF) with 'Kenland'. Dr. Taylor continued with multiple generations of recurrent selection including 2,4-D applications each cycle. After his death, UK researchers continued with modifications to the original protocol and are now in final stages of testing the most advanced line UK2022. To validate tolerance a trial was planted spring 2023 and sprayed late April 2024 (2nd production yr). Injury ratings (0-9 scale) two wks post-application at 1lb/A ae were 2.0, 4.0, and 4.5 and at 2lb/A ae 3.5, 5.25, and 6.0 for UK 2022, 'Q Medium Red', and Kenland, respectively. Regrowth injury ratings 7 wks post-application (4 wks after harvest) were 0.0, 0.75, 5.25 at 1lb/A ae and 0.75, 0.75, 6.0 at 2lb/A ae for UK 2022, Q Red, and Kenland. Cumulative forage yield 2024 (3 harvests) for unsprayed checks was 5.1, 4.4, 5.6 tons/A for UK 2022, Q Red, and Kenland. The yield for these entries at 1lb/A ae was 4.6, 3.3, 3.3 and at 2lb/A ae 4.2, 3.0, 2.8 tons/A. Lower yields for Q Red in KY was likely related to it being an early maturing variety and reduced 2nd year persistence. Tara Lewis determined potential mode of tolerance in these breeding lines. Using an earlier 2,4 D tolerant line (UK2014), she showed less parent 2,4-D and more 2,4-D metabolites leaf, shoot and root sections of UK2014 at 8, 24, 48 and 72 hrs post-treatment compared to Kenland indicating enhanced 2,4-D metabolism is the likely basis for tolerance. In conclusion, UK 2022 shows promise for low-rate applications of 2,4 D with less injury and better recovery. Further testing to validate 2,4 D tolerance in UK 2022 is planned before a potential variety release.

[1]S. Ray Smith, Univ. of KY (raysmith1@uky.edu); remaining authors Univ. of Kentucky.

FORAGE AND COVER CROP BREEDING FOR THE SOUTHEASTERN US: UF-NFREC PROGRAM OBJECTIVES

C. de Souza[1], J. Dubeux and A. Blount

The Southern Coastal Plain of the southeastern U.S. offers opportunities and challenges for forages and cover crops. While the mild climate supports diverse species, low-fertility sandy soils, temperate winters, and tropical summers necessitate resilient forages to withstand environmental stresses and pests. The forage breeding program at the University of Florida's North Florida Research and Education Center develops species with improved forage yield, seasonal distribution, nutritive value, and resilience, thereby supporting sustainable livestock systems in the Southeastern U.S. The program focuses on: 1) Drought-tolerant perennial grasses: breeding tetraploid and diploid bahiagrass lines with improved forage and seed production traits, disease resistance, and adaptability to low-fertility soils; 2) Forage legumes for mixed pastures and cover crops (seed-propagated pintoi peanut, rhizoma peanut for grazing persistence, sun hemp with improved seed production, and common vetch for fall establishment and early forage production in mixed plantings with small grains or overseeding in perennial grasses, and a perennial forage legume rich in condensed tannins that can survive winters in North Florida; 3) Annual grasses (breeding black oats, and small grains for early forage and disease resistance; 4) Seed production and variety adoption: optimizing seed production practices for crimson clover, sunn hemp, and vetch while collaborating with regional universities to strengthen the Southeastern seed industry. With these objectives, this program aims to deliver resilient, high-performing forage varieties, contributing to sustainable agriculture in the region.

[1]Assistant Professor Forage Breeding, University of Florida, North Florida Research and Education Center – Marianna, FL (c.lopesdesouza@ufl.edu).

HERBAGE AND LIVESTOCK RESPONSES IN CONTRASTING GRAZING SYSTEMS

K.R. Trumpp, J.C.B. Dubeux Jr., J. Portuguez Acuña, K. Oduor, I. Lima Bretas, L.M.D. Queiroz, M. Araujo Bernardini, M. Ruiz-Moreno, L. Garcia, L.E. Sollenberger, N. DiLorenzo, and J.M.B. Vendramini[1]

Rhizoma peanut (*Arachis glabrata* Benth) is a N-fixing, warm-season legume with the potential to reduce external nitrogen (N) inputs in tropical and subtropical grazing systems. We evaluated the effects of different N fertilization levels and legume inclusions into bahiagrass (*Paspalum notatum* Flügge) pastures during the cool and warm seasons on herbage and animal responses. The Grass+N treatment consisted of N-fertilized bahiagrass during the summer and cool-season grasses; each season received 112 kg N/ha/yr. The Grass+CL consisted of bahiagrass during the summer with no N fertilizer, a mixture of cool-season grasses, and a blend of clovers receiving 34 kg N/ha/yr. The Grass+RP consisted of a bahiagrass and rhizoma peanut mixture during the summer with no N fertilizer and a mixture similar to that of cool-season grasses and clovers of the Grass+CL. Legumes extended the length of the grazing season, and the BNF in the Grass+RP was around 103 kg N/ha/yr. During the cool season, there was a greater stocking rate for the Grass+N, averaging 3.6 compared to 2.4 and 2.8 AU/ha for the Grass+CL and Grass+RP, respectively. Including rhizoma peanut had similar yearly animal performance as N-fertilized grass systems while reducing N fertilizer inputs by 85%. Liveweight gain on Grass+N, Grass+CL, and Grass+RP systems averaged 764, 520 and 630 kg/ha, respectively.

[1]PhD Student (ktrumpp@ufl.edu); Professor Forage Agronomy (dubeux@ufl.edu); MS (portuguez.javier@ufl.edu); PhD Student (kenneth.oduor@ufl.edu); Post-doctoral associate (ig.limabretas@ufl.edu); PhD Student (ldantasqueiroz@ufl.edu); MS Student (m.araujobernardi@ufl.edu); Biological Scientist (ruizm001@ufl.edu); Research Coordinator (lizagarcia@ufl.edu); Professor Animal Nutrition (ndilorezno@ufl.edu) - North Florida Research and Education Center, Marianna, FL; Professor Forage Agronomy, Gainesville, FL (lesollen@ufl.edu); Professor Forage Agronomy, Range Cattle Research and Education Center Ona, FL (jv@ufl.edu).

EVALUATION OF BIOMASS AND GROWTH DYNAMICS OF ALFALFA UNDER LIMITED IRRIGATION

J.Y. Kim, S. Dey, G. Jha, and D.H. Min[1]

Alfalfa (*Medicago sativa* L.) is valued for its high biomass and adaptability to Kansas. Given its water-dependency and the depletion of resources such as the Ogallala Aquifer, optimizing irrigation is crucial for sustainable production. This study evaluates the effects of alfalfa on growth characteristics and production under limited irrigation in the controlled environment. Alfalfa variety was an L442RR (Fall dormancy 4.4). The PVC pots used were 50 x 20 cm in length and diameter. The greenhouse was maintained at a temperature of 24/18°C in day/night, with continuous lighting. Harvesting was conducted four times at 30-day interval. Treatments were based on irrigation level corresponding to field capacity (FC, %) by 55 (FC55), 70 (FC70), 85 (FC85), and 100 (FC100), respectively. The result showed that higher FC85 and FC100 promote above- and below-ground biomass production and the increased root/shoot and leaf-to-stem ratios at lower FC55 suggested that alfalfa prioritizes root growth and leaf area expansion under water-limited conditions. This means adaptation under harsh conditions can increase water uptake and photosynthetic activity during drought stress, but it may come at the cost of reduced overall biomass productivity. Interestingly, in NDVI, higher FC85 and FC100 were found, indicating that it can monitor alfalfa growth and biomass under water-limited conditions. Future research would be focused on how different alfalfa varieties respond in the field under the different irrigation rates to maximize both alfalfa yield and quality.

[1]Research scholar, Department of Agronomy, Kansas State University, Manhattan, KS (jiyung@ksu.edu); Ph D student, Department of Agronomy, Kansas State University, Manhattan, KS (sdey@ksu.edu); Assistant Professor, Department of Agronomy, Kansas State University, Manhattan, KS (gjha@ksu.edu); Professor, Department of Agronomy, Kansas State University, Manhattan, KS (dmin@ksu.edu).

The Effect of Bermudagrass Cultivar on the Energetics of Beef Heifers

Abbigail R. Hines, [1] Chandler Kyle, [1] Diva N. Rigney, [1] Anna G. Underwood, [1] Werner G. Bergen,[1] M. Kimberly Mullenix,[1] S. Leanne Dillard,[1] Todd R. Callaway,[2] and W. Brandon Smith[1]

[1]Department of Animal Sciences, Auburn University, Auburn, AL 36849
[2]Department of Animal and Dairy Sciences, University of Georgia, Athens, GA 30602

Although much conducted research in the last 80 years has sought to genetically improve bermudagrass (Cynodon dactylon [L.] Pers.) cultivars for traits such as yield, digestibility, and nutritive value, there is limited research comparing cultivars under *in vivo* conditions of the grazing animal. Therefore, the objective of this study was to evaluate the energetics of bermudagrass cultivars from heifers consuming four bermudagrass cultivars commonly found in the Southeastern U.S. In an incomplete block design, ruminally-fistulated heifers (n = 3) were randomly allocated one of four bermudagrass cultivars (Coastal [COS], Russell [RUS], Tifton 44 [T44], or Tifton 85 [T85]) for four 30-d in vivo periods (21-d adaptation and 9-d collection). From d 14 to d 30 of each period, hay and ort samples from each heifer were collected. Urine and total fecal samples were collected from d 21 to 25. Hay, ort, and fecal samples were ground and composited based on an assumed 48-h passage rate. All samples were assayed for gross energy [GE], digestible energy [DE], metabolizable energy [ME], and net energy [NE]. Data were analyzed using SAS v. 9.4. There was no effect of treatment ($P \geq 0.13$) for GE of intake (4.2 Mcal/kg), GE output (4.3 Mcal/kg), dietary concentrations of DE (2.2 Mcal/kg), ME (1.8 Mcal/kg), or NEma (0.9 Mcal/kg). Maintenance requirements were met by the bermudagrass cultivars for beef heifers evaluated in this study. Therefore, cultivar differences may not impact the available digestive and metabolic energy to beef cattle in grazing bermudagrass cultivars of similar maturity.

For title and summary submission:

The objective of this study was to evaluate the energetics of bermudagrass cultivars from heifers consuming four bermudagrass cultivars commonly found in the Southeastern U.S. From a metabolism trial using 3 beef heifers, hay, ort, and fecal samples were ground and composited based on an assumed 48-h passage rate. All samples were assayed for gross energy [GE], digestible energy [DE], metabolizable energy [ME], and net energy [NE].

INTEGRATED CROP-LIVESTOCK SYSTEMS IN MISSISSIPPI

B. Rushing[1], J.G. Maples[2], K.M. Harvey[3], and J.C. Lyles[4]

The integration of grazing cover crops in combination with soybean production has the potential to increase total farm revenue. Our objectives were to determine the impact grazing had on subsequent soybean production and the economic implications of this practice. A field trial was conducted at the Coastal Plain Branch Experiment Station (CPBES) in Newton, MS and the Prairie Research Unit (PRU) in Prairie, MS from 2021 to 2023 to compare three cropping systems on two distinct soil types. Cropping systems were conventional soybean (CS); no-till soybean + cereal rye cover crop (CC); and no-till soybean + grazed cereal rye cover crop (GC). Analysis was separated by location. Cover crop, soybean production, animal performance, and economic analysis were evaluated for each treatment. Soybean grain yield varied by treatment; GC (54.6 bu ac^{-1}) was greater than CS (52.3 bu ac^{-1}) at CPBES. At PRU, CS (68.5 bu ac^{-1}) had greater soybean yield than all other treatments. Cover crop forage mass (FM) was 5,077 lb ac^{-1} at CPBES, compared to 3,094 lb ac^{-1} at PRU, resulting in subsequent cattle revenue of \$593.64 and \$160.29 ac^{-1} for CPBES and PRU, respectively. Soybean revenue was greatest for GC at CPBES (\$691.78 ac^{-1}) and CS at PRU (\$867.89 ac^{-1}). Net returns above production costs were greatest for GC at CPBES (\$811.59 ac^{-1}) and CS at PRU (\$528.58 ac^{-1}). Findings suggest grazing cereal rye cover crop has the potential to increase net returns in a no-till soybean system on coarse textured soils; but reduces soybean grain yield on heavy, poorly drained sites.

[1]Associate Professor; Mississippi State University; Coastal Plain Branch Experiment Station.
[2]Associate Professor; Mississippi State University; Department of Agricultural Economics.
[3]Assistant Professor; Mississippi State University; North MS Research and Extension Center.
[4]Research Associate II; Mississippi State University; Coastal Plain Branch Experiment Station.

NEED FOR PHOSPHORUS SUPPLEMENTS FOR BEEF HEIFERS ON PASTURES

C.P. Bagley[1], Robert Palfreyman[2], and Jennifer Crabtree[3]

Beef cattle and calves are part of the farm enterprise in just over 70% of all farms in the US. Most beef herds are around 25 head in number, meaning superior management techniques are not always employed. Mineral supplements are routinely fed to beef cattle herds, with costs per ton ranging from just over $100 to $1,000/ ton. Much of the cost variance is based on the level of supplemental P in the mineral mixture, which is a costly mineral and is usually supplemented at 3 – 6% of the mineral supplement. Distiller's Dried Grains (DDGS) are a co-product of the ethanol industry and is a newer protein supplement containing a relatively high level of available P. In this study, crossbred 14-month-old beef heifers (n=36) were rotated on 3 different mineral supplement treatments containing different levels of phosphorus or different levels of enzymes. Each treatment was fed 28 d and treatments rotated with three groups of heifers receiving all three treatments. Results showed no difference in the Group vs Treatment in P content for both blood and feces, and no differences in animal gains were detected. These results would indicate that in situations where producers use DDGS as a protein supplement, there is less of a need for high levels of supplemental P in the mineral premix, resulting in a reduction in feed costs.

[1]Assoc. Vice Chancellor, Research (interim) and Director, Beef Research Institute, Southern University; [2]Natural Resource Conservation Service Specialist, Tennessee; [3]Ph.D. Graduate Assistant, Auburn University.

GROWTH AND ECONOMICS OF HEIFERS RAISED ON PASTURE COMPARED TO HEIFERS RAISED IN CONFINEMENT

C. H. P. Camisa Nova[1], K.F. Kalscheur[2], and D. Jaramillo[3]

Raising dairy heifers on pasture is a tool for reducing costs associated with replacement heifers. This study evaluated the performance of dairy heifers raised either on pasture or in confinement on potential long-term effects on lactation and on economic implications under both scenarios. In this study, 18 heifers each were raised on either cool-season grass pastures or in a free-stall barn fed a total mixed ration (TMR) diet. After the first calving, all animals were submitted to conventional management for lactating dairy cows. Age at calving was similar for both treatment groups, which averaged 24.3 months. Body weights were similar at calving; pasture-raised heifers were 1,310 lbs. and confinement-raised heifers were 1,355 lbs. Raising heifers on pasture during the first grazing season reduced costs by 27% compared with heifers raised in confinement (US$ 142.47 vs US$ 194.48). During the second grazing season, costs associated with raising heifers on pasture were reduced by 58% compared to heifers raised in confinement (US$ 97.43 vs US$ 234.18). During the first lactation, cows who were previously raised on pasture produced more milk than those previously raised in confinement (27,199 and 24,487 lbs., respectively). Consequently, a greater average income over feed costs was found for cows previously raised on pasture compared to those raised in confinement (US$ 12.80/d and US$ 11.04/d, respectively). Thus, economic advantages are evident when heifers are raised on pasture without deleterious effects on growth and performance.

[1]PhD Candidate, University of Wisconsin – Madison, Madison, WI (camisanova@wisc.edu); [2]Research Animal Scientist, USDA-ARS, U.S. Dairy Forage Research Center, Madison, WI (kenneth.kalscheur@usda.gov); [3]Research Animal Scientist, USDA-ARS, Environmentally Integrated Dairy Management Research, Marshfield, WI (david.jaramillo@usda.gov).

HOW DAIRY PRODUCERS WILL FEED COWS IN THE FUTURE

D.W. Hancock[1]

Dairy producers continue to deal with a feed cost:milk price ratio that continues to increase. This is creating pressure to increase labor use efficiency. Further, dairy producers must cope with economic and societal pressures to be more efficient users of nutrients, improve the health and well-being of their animals, and decrease emissions of methane, nitrous oxide, and ammonia. These pressures can partially be addressed with robots that automatically push-up feed, automated feeding systems (AFS), and automated milking systems (AMS), and is why the dairy industry is seeing a growing prevalence of these systems. Decision Support Tools and use of artificial intelligence likely will drive nutritional decisions to an ever-finer management resolution (i.e., from herd to nutritional group to individual animal feeding) in the near future. A review of literature shows indications that automated push-ups can increase milk production, AFS can improve feed efficiency and animal health, and AMS could enable supplementation rations and rates tailored to more nutritional groups or even an individual cow that could further improve all these metrics of dairy performance. Additionally, these systems could provide the dairy producer with systems that simultaneously measure and monitor individual, group, and herd performance. With improved ability to measure and monitor, dairy producers should have greater opportunities to manage for decreased emissions and increased labor efficiency, nutrient use efficiency, and animal health. Hopefully, these innovations would be economical and enable dairy producers to utilize these innovations in the face of increased feed cost:milk price pressure.

[1]Center Director, U.S. Dairy Forage Research Center, USDA-Agricultural Research Service, 1925 Linden Dr., Madison, WI 53706 (Dennis.Hancock@usda.gov; 608-215-0223).

EFFECT OF COVER CROPS AND GRAZING MANAGEMENT ON SOIL MICROBIAL ACTIVITY INDICATORS IN CONTRASTING CROP SYSTEMS

C. Garcia, J. Dubeux, E. Santos, G. Fracetto, F. Fracetto, and L. Silva[1]

Abstract

Soil microbial activity is vital for soil health and fertility, aiding nutrient cycling and organic matter decomposition. Cover crops enhance soil organic matter and nutrients, while grazing management affects soil disturbance and residue levels, influencing microbial communities. This study assessed the impact of cover crops and grazing management on soil microbial activity indicators in crop rotation (SBR) and integrated crop-livestock systems (ICL). The experiment was conducted at the North Florida Research and Education Center, Marianna, Florida. The study was under a randomized complete block design with three replications with nine treatments combining row-crop systems (cotton), cover crops (oat and ryegrass), and grazing management at varying post-grazing residue levels (225, 675 and 1125 kg DM^{-1}). Treatments consisted of: T1: Fallow; T2: Cover crop (no grazing, no nitrogen); T3: Cover crop (no grazing, 80 lb N/acre); T4: SBR - overgrazing; T5: SBR - moderate grazing; T6: SBR - lenient grazing; T7: ICL - overgrazing; T8: ICL - moderate grazing; T9: ICL - lenient grazing. Lenient grazing (T6 and T9) and nitrogen application (T4) resulted in higher (P=0.01) microbial biomass carbon (MBC) (493, 442, and 440 respectively vs. 146 µg C g^{-1} soil in T1). No differences were observed in basal respiration and metabolic quotient averaging 20.70 mg C-CO_2 kg^{-1} day^{-1} and 17.88 mg qCO_2 g^{-1} day^{-1}, respectively . Greater MBC and metabolic quotient were observed in 2019 over 2020 (358 vs. 310 µg C g^{-1} soil and 21 vs. 15 mg qCO_2 g^{-1} day^{-1}, respectively). These results suggest that post-grazing residue management can increase microbial biomass carbon, while seasonal factors influence soil microbial activity, contributing to more efficient, sustainable agriculture adapted to climate change.

Introduction: Soil microbial activity is essential for nutrient cycling and organic matter decomposition, directly influencing soil health and fertility. Practices such as cover crops and grazing management can enhance soil quality by increasing organic matter and optimizing soil structure (Garcia et al., 2025). Lenient grazing management, for example, promotes microbial biomass through reduced compaction and greater organic inputs (Six et al., 2002; Luo et al., 2010).

This study aimed to evaluate the effects of cover crops and grazing intensities on microbial activity indicators, such as microbial biomass carbon (MBC), soil basal respiration (SBRe), and the metabolic quotient (qCO_2), in integrated crop-livestock (ICL) and crop rotation systems (SBR) to provide insights into how these practices can improve soil health and contribute to sustainable agricultural systems.

---

[1]Postdoctoral Fellow, Clemson University, Blackville, SC (cvelaga@clemson.edu); Associate Professor, University of Florida, USA (dubeux@ufl.edu); Assistant Professor, University of Alberta, Canada (ericksantos@alberta.ca); Adjunct Professor, Federal Rural University of

Pernambuco, Brazil (curyfelipe@hotmail.com); Position, Postdoctoral Fellow, Federal Rural University of Pernambuco, Brazil (mercia.vfsantos@ufrpe.edu.br); Assistant Professor, Clemson University, Blackville, SC (lseveri@clemson.edu).

Materials and Methods: The study was conducted at the North Florida Research and Education Center (NFREC) in Marianna, Florida (30°46′35″ N and 85°14′17″ W; 51 m above sea level). Soils are classified as typical of the Jackson Blue Spring and Chipola River basins. The study evaluated the effects of in-crop rotation (SBR) and integrated crop-livestock systems (ICL) on soil microbial activity indicators, including soil basal respiration and the metabolic quotient. The SBR rotation included bahiagrass (*Paspalum notatum* F.; 1st and 2nd years) and peanut (*Arachis pintoi* K.; 3rd year), while the ICL rotation featured cotton (*Gossypium hirsutum* L.; 1st and 2nd years) and peanut (3rd year). These indicators were measured only during the first and second years of the experiment and not during the third year when peanuts were planted. Nine treatments were assessed in a randomized complete block design with three replications. Treatments included: T1: Fallow, T2: Cover crop (no grazing, no nitrogen), T3: Cover crop (no grazing, 80 lb N/acre), T4: SBR – overgrazing +80 lb N/acre and 225 kg DM^{-1} $acre^{-1}$ of post-grazing residual stubble mass, T5: SBR - moderate grazing +80 lb N/acre and 675 kg DM^{-1} $acre^{-1}$ of post-grazing residual stubble mass, T6: SBR - lenient grazing +80 lb N/acre and 1125 kg DM^{-1} $acre^{-1}$ of post-grazing residual stubble mass, T7: ICL – overgrazing +80 lb N/acre and 225 kg DM^{-1} $acre^{-1}$ of post-grazing residual stubble mass, T8: ICL - moderate grazing +80 lb N/acre and 675 kg DM^{-1} $acre^{-1}$ of post-grazing residual stubble mass, T9: ICL - lenient grazing +80 lb N/acre and 1125 kg DM^{-1} $acre^{-1}$ of post-grazing residual stubble mass. Soil samples were collected at 0–10 cm, refrigerated, and prepared for analysis. Microbial biomass carbon (MBC) was determined using the fumigation-extraction method described by Vance et al. (1987) and soil basal respiration (SBRe) and metabolic quotient (qCO_2) were determined using the method described by Anderson and Domsch (1986). Data were analyzed utilizing the PROC MIXED from SAS (SAS Institute), and LSMEANS were compared using PDIFF option, adjusted by Tukey's test ($p < 0.05$).

Results and Discussion: Soil microbial activity indicators differed among treatments, confirming the effect of management practices on soil health. The higher MBC values ($P = 0.01$) were observed in treatments 6, 9, and 4, with 493, 442, and 440 μg g^{-1} soil respectively, while the lowest value was reported in treatment 1 (146 μg g^{-1} soil; Table 1). These findings align with studies showing that less intensive management practices, such as lenient grazing management, promote microbial biomass accumulation due to increased organic matter input and reduced soil compaction (Six et al., 2002; Luo et al., 2010). Although SBRe exhibited higher values in treatments 9 and 7, respectively, 26 and 24 mg C-CO_2 kg^{-1} day^{-1}, the differences were not statistically significant ($P = 0.07$). This trend may be attributed to increased organic matter mineralization in areas subjected to higher grazing pressure, a pattern previously observed by Shan et al. (2011) in pasture systems. On the other hand, the qCO_2, an indicator of microbial biomass efficiency, also showed no significant differences ($P = 0.69$) (Table 1). The highest values in treatment 6 (24 mg g^{-1} day^{-1}) suggest increased metabolic stress in response to organic matter input, as reported in studies by Tisdall et al. (2020).

Interannual analysis revealed significant differences for MBC and qCO_2, with higher values in 2019 (Table 2). This pattern may be linked to more favorable climatic conditions, such as greater moisture availability and moderate temperatures, which enhance microbial activity, as reported by Butcher et al. (2020). In 2020, the increase in SBR could reflect greater decomposition of accumulated organic matter, while the lower qCO_2 suggests reduced microbial efficiency, possibly due to environmental stress or lower-quality carbon inputs, a phenomenon described by Kuzyakov et al. (2000). From a practical perspective, the results highlight the importance of moderate and lenient management practices for promoting soil health. Integrated crop-livestock systems (ICL) and crop-rotation (SBR) systems can benefit from strategies that mitigate the negative impacts of overgrazing and soil compaction, as demonstrated by Franzluebbers (2007). Moreover, interannual variability underscores the need for resilient management practices that consider climatic conditions and continuous monitoring of biological indicators.

Thus, this study contributes to understanding the interactions between management practices, soil microbiota, and sustainability. It reinforces that less intensive practices and the use of integrated systems can enhance the resilience and productivity of agricultural systems, aligning with the principles of sustainable agriculture and mitigating environmental impacts. In conclusion, these results suggest that post-grazing residue management can increase microbial biomass carbon. At the same time, seasonal factors influence soil microbial activity, contributing to more efficient, sustainable agriculture adapted to climate change.

Literature cited:

Six, J., Conant, R. T., Paul, E. A., & Paustian, K. (2002). Stabilization mechanisms of soil organic matter: implications for C-saturation of soils.

Plant and soil, 241, 155-176. Luo, Z., Wang, E., & Sun, O. J. (2010). Can no-tillage stimulate carbon sequestration in agricultural soils? A meta-analysis of paired experiments. Agriculture, ecosystems & environment, 139(1-2), 224-231.

Shan, Y., Chen, D., Guan, X., Zheng, S., Chen, H., Wang, M., & Bai, Y. (2011). Seasonally dependent impacts of grazing on soil nitrogen mineralization and linkages to ecosystem functioning in Inner Mongolia grassland. Soil Biology and Biochemistry, 43(9), 1943-1954.

Tisdall, J. M. (2020). Formation of soil aggregates and accumulation of soil organic matter. In Structure and organic matter storage in agricultural soils (pp. 57-96). CRC Press.

Butcher, K. R., Nasto, M. K., Norton, J. M., & Stark, J. M. (2020). Physical mechanisms for soil moisture effects on microbial carbon-use efficiency in a sandy loam soil in the western United States. Soil Biology and Biochemistry, 150, 107969.

Kuzyakov, Y., Friedel, J. K., & Stahr, K. (2000). Review of mechanisms and quantification of priming effects. Soil Biology and Biochemistry, 32(11-12), 1485-1498.

Franzluebbers, A. J. (2007, August). Soil physical aspects of integrated crop-livestock systems. In Proc. Int. Symp. Crop-Livestock Syst (pp. 13-15).

Luo, Z., Wang, E., & Sun, O. J. (2010). Can no-tillage stimulate carbon sequestration in agricultural soils? A meta-analysis of paired experiments. Agriculture, ecosystems & environment, 139(1-2), 224-231.

Table 1. Soil microbial activity indicators under cover crops and grazing management.

Treatments	MBC ($\mu g\ g^{-1}.\ solo^{-1}$)	SBRe ($mg.C\text{-}CO_2.kg^{-1}.dia^{-1}$)	qCO_2 ($mg\ g^{-1}.\ dia^{-1}$)
1	146 b	13	14
2	275 ab	21	15
3	218 ab	19	13
4	440 a	24	22
5	382 ab	23	20
6	493 a	21	24
7	346 ab	24	16
8	260 ab	16	19
9	442 a	26	19
P value	0.01	0.08	0.69
SE	109	4	3

*MBC: microbial biomass carbon, SBRe: soil basal respiration, qCO_2: metabolic quotient, SE: standard error. Values followed by different a lowercase letter within a column indicate treatments are different at the 5% probability level.

Table 2. Soil microbial activity indicators under cover crops and grazing management in 2019 and 2020.

Year	CBM ($\mu g\ g^{-1}$ $solo^{-1}$)	SBRe ($mg.C\text{-}CO_2.kg^{-1}.dia^{-1}$)	qCO_2 ($mg\ g^{-1}\ dia^{-1}$)
2019	358 a	19	21 a
2020	310 b	23	15 b
P value	0.01	0.05	0.03
SE	24	2	3

*MBC: microbial biomass carbon, SBRe: soil basal respiration, qCO_2: metabolic quotient, SE: standard error. Values followed by different a lowercase letter within a column indicate treatments are different at the 5% probability level.

EFFECTS OF GRAZING COOL-SEASON ANNUAL MIXTURES ON CROP PERFORMANCE AND SOIL HEALTH

Suazo, Pereira, Mullenix, Smith, and Dillard[1]

Grazing cool-season cover crops offers an environmentally sustainable approach to extending the grazing season and reducing the ecological footprint of traditional row-crop farming systems. This study investigated the effects of various cover crop mixtures on cotton (*Gossypium hirsutum*)-peanut (*Arachis hypogaea*) rotations. The research evaluated soil characteristics, forage biomass, and nutritional composition. Conducted at the Wiregrass Research and Extension Center in Headland, AL, the trial included 18 paddocks planted with grass ['Coker' oat (*Avena sativa*)], grass + legume ['Dixie' crimson clover (*Trifolium incarnatum*)], or grass + legume + brassica [('Turnip' brassica (*Brassica rapa* subsp. *rapa*)]. These paddocks were randomly assigned to grazed or un-grazed conditions (n = 3). Yearling beef steers (266 ± 44 kg) grazed using a put-and-take system to maintain a forage allowance of 1 kg DM/kg body weight. Biweekly forage samples were collected to measure biomass, botanical composition, and nutritional value. Grazing lasted 65 days, from January 17 to March 22, 2023. Steer body weights were recorded at the start and end of the grazing period to calculate average daily gain. Statistical analysis revealed no significant interaction between treatments and grazing ($p = 0.4604$) or differences among treatments ($p = 0.6696$). However, grazed paddocks exhibited higher crude protein content than ungrazed paddocks ($p = 0.0173$). These findings highlight the potential benefits of cover crops for improving forage quality and nutrient cycling, though their full impact on soil health and sustainability requires further exploration. Integrating cover crops into cropping systems supports agricultural resilience and promotes environmental sustainability.

[1]Department of Animal Sciences, Auburn University, AL.

USING PERENNIAL FORAGES TO REDUCE INPUT COSTS OF INTEGRATED FORAGE-ROW CROP PRODUCTION

Eric Billman[1]

Development of an integrated forage-cotton crop system using cool-season perennial forages could allow cotton and cattle production on the same unit of land, thus increasing sustainable intensification. This study evaluated input reduction and economic benefits of incorporating a cool-season perennial forage groundcover system into cotton in the southeastern U.S. Coastal Plain from October 2022 – October 2024. Six treatments were established in fall 2022: 1) weedy fallow, 2) annual ryegrass monoculture, 3) tall fescue monoculture, 4) 1:1 red clover and white clover mixture, 5) a 2:1:1 annual ryegrass, red clover, and white clover mixture, and 6) a 2:1:1 tall fescue, red clover, and white clover mixture. Forages were harvested during spring each year, then strip tilled and intercropped with cotton. The fallow and annual ryegrass treatments were managed conventionally with full pesticide application, but all other treatments received only a single in-row banding of glyphosate post-cotton emergence. Economic inputs and profit were determined from both the forage and cotton components and compared among treatments. Despite $10 - $49/acre lower input costs where growing perennial forages, lint yield reductions where perennial forages were grown reduced profit margins even when not using pesticides. However, cotton grown alongside the red and white clover mixture had similar profitability ($351/acre) to the fallow ($492/acre) and annual ryegrass controls ($396/acre). When combined with cattle gains from forage, profitability of a forage-cotton cropping system would likely exceed a conventional cotton monoculture using conventional pesticide inputs.

[1]Research Agronomist, USDA-ARS, Coastal Plains Soil, Water, and Plant Research Center, Florence, SC (eric.billman@usda.gov).

COMPARATIVE PERFORMANCE OF LATE MATURING ANNUAL CLOVERS UNDER ANNUAL RYEGRASS GRAZING

R. Lemus[1] and J. White[2]

Increases in the cost of fertilizer has sparked the need for legumes that can increase sustainability of cattle production during the winter months in the southern USA. This study compared production of annual ryegrass monoculture or a ryegrass/clover system on animal performance during two fall/winter grazing seasons. Treatments consisted of annual ryegrass monoculture (RG), annual ryegrass + balansa clover (RBA), annual ryegrass + berseem clover (RBE), and annual ryegrass + Persian clover (RPR) planted in a prepared seedbed. Ryegrass was planted 25 and 20 lb PLS/ac as monoculture or clover mix, respectively and clover species at 8 lb PLS/ac. Ryegrass monoculture was fertilized at 100 lb N/ac as urea (46-0-0) at 3-in height after germination and after the first grazing cycle. Grass/clover mixes received 25 lb N ac^{-1} at three inches in height after germination. Grazing began at 12 inches height. Cattle were weighed pre- and post-grazing for each grazing event to calculate average daily gain (ADG) and gain per acre (GPA). Cattle were removed when forage biomass reached 700 lb DM/ac and allowed to recover until they reached 12 inches in height. Harvested samples were analyzed for CP, NDF, ADF, TDN, water-soluble carbohydrates (WSC), and *in vitro* true DM digestibility (IVTDMD). There were no differences in ADG or GPA for the treatments. The ADG ranged from 1.7 (RPR) to 2.3 (RBE) lb d^{-1} and gains ranged from 331 (RPR) to 403 (RG) lbs of beef ac^{-1}. There were treatment differences in CP ($P = 0.0266$), and IVTDMD. The cost of fertilizer might contribute to a higher cost of gains during the study compared to introducing late-maturing annual clovers.

[1]Professor and Extension Forage Specialist, Mississippi State University, MS (Rocky.Lemus@msstate.edu); [2]Forage Variety Testing Manager, Mississippi State University, MS (JWhite@pss.msstate.edu).

A COMPARISON BETWEEN PERENNIAL AND PERENNIAL/ANNUAL COMBINATION FORAGE SYSTEMS UNDER MIXED-SPECIES GRAZING

A.M. Grev[1], J.W. Semler[1], and M.E. Townsend[1]

Annual forages can be used to increase forage availability across the growing season; however, their continued establishment requires additional time and financial inputs. To explore this tradeoff, a study was started at the Western Maryland Research and Education Center to compare a traditional, perennial-based grazing system against a perennial/annual combination grazing system on the basis of forage production and quality, livestock performance, soil health, and economic feasibility. In 2023, 30 lambs and 19 calves were blocked by age, birth type (lambs), and bodyweight and divided into two mixed-species groups. Animals in the perennial-only forage system (PER) were rotationally grazed across 13 acres of cool-season perennial pasture. Animals in the perennial/annual combination forage system (ANN) were rotationally grazed across 13 acres of both perennial and annual pastures. Both treatment groups had ad libitum access to mineral and were rotated to a new paddock every 2-5 days depending on forage availability. Animals were measured [BW, BCS, FAMACHA© (lambs only)] and fecal samples were collected on a bi-weekly basis throughout the grazing season (May-Sept). Data was analyzed using the MIXED procedure of SAS ($P \leq 0.05$). Initial BW, BCS, and FAMACHA© scores did not differ between treatment groups. Lamb final BW and FAMACHA© scores did not differ, but the ANN lambs had higher ADG (0.26 lb/d) and final BCS (4.0) compared to the PER lambs (0.23 lb/d and BCS 3.75). Calf final BCS and ADG did not differ, but final BW tended to be greater for ANN calves (673 lb) compared to PER calves (601 lb). Initial results indicate potential advantages for the ANN system; the study will be repeated in 2024-2025.

[1]University of Maryland Extension, WMREC, Keedysville, MD (agrev@umd.edu).

EFFECTS OF STOCKING PERIOD ON PASTURE SPECIES COMPOSITION, FORAGE PRODUCTION, AND ANIMAL PERFORMANCE

David M. Jaramillo[1], Arthur Franco[2], Marta Kohmann[2], and Mark Renz[2]

Increasing the length of stocking period has been identified as a management practice for improving productivity in rotationally stocked grazing systems. Quantifying differences from animal and pasture performance will provide key data currently missing from the literature and allow for evidence-based recommendations for grazing farmers. This study evaluated three stocking periods (1-, 4-, and 8-d; 32-d rest period across treatments) in a randomized complete block design with three replicates. Pastures were composed of perennial cool-season grasses [mostly orchardgrass (*Dactylis glomerata*) and smooth brome, (*Bromus inermis*), and legumes (clovers; *Trifolium* spp.). Weeds present were dandelion (*Taraxacum officinale*), Canada thistle (*Cirsium arvense*), and yellow foxtail (*Setaria pumila*)]. Holstein-Angus heifers (~10-12 mo) were used under variable stocking rates. Herbage mass and nutritive value were assessed every 14-d. Treatment had no significant effect on herbage or animal responses. Herbage mass showed a bi-modal distribution across the season. Nutritive value was not affected by treatment. The average crude protein and in vitro digestible organic matter concentrations across the season and treatments were 12 and 52%, respectively. Average daily gain did not differ across treatments, and was 0.44 kg/head/d, which was below the expected average, likely due to drought in 2023. Weed biomass was similar across treatments (avg. 4.63 g DM/m). Stocking periods did not affect pasture or animal performance. Further long-term research to assess cumulative treatment effects and animal supplementation would be valuable to inform management practices.

[1]USDA-ARS, U.S. Dairy Forage Research Center, Marshfield, WI (david.jaramillo@usda.gov);
[2]Plant and Agroecosystem Sciences Department, University of Wisconsin, Madison, WI.

PERCEPTIONS OF ESTABLISHMENT AND ENVIRONMENTAL IMPACTS OF SILVOPASTURE BY PRODUCERS IN THE SOUTHEASTERN UNITED STATES

H. Malcomson, G. Agunbiade, D. Sahoo, L. O'Halloran, and L.S. Silva[1]

Silvopasture is a climate-resilient grazing system that incorporates timber, cattle, and forages, allowing producers to improve their property's ecosystem and potentially improve carbon sequestration, while effectively growing cattle and timber. This study involved a 14-question survey distributed to producers in South Carolina, North Carolina, Alabama, Virginia, Mississippi, and Florida. The survey was written to gauge producer opinions on silvopasture, specifically barriers to establishment, forage and grazing management, environmental concerns on farm, and overall interest in learning more about silvopasture logistics and benefits. When asked to rate their interest out of ten, producers averaged a score above seven for the environmental benefits of silvopasture, diversification of income through silvopasture establishment, cost to introduce silvopasture and upkeep, and generating carbon credits for future carbon markets. Potential carbon markets were also of high interest to the majority of farmers, with the average interest score across all producers being 3.6/5. Other environmental concerns, specifically water and soil quality, were of some interest to producers (avg: 2.96/5 and 2.97/5, respectively), although over 70% of producers stated that they had never tested their groundwater quality. Overall producer interest in silvopasture establishment and its environmental benefits centered around carbon sequestration for carbon credits and diversification of income, with possible room for educational resources regarding water and soil quality testing practices.

[1]Ph.D. student, Clemson Univ., Clemson, SC (hmalcom@clemson.edu); Ph.D. student, Clemson Univ., Clemson, SC (gagunbi@g.clemson.edu); Assistant Professor, Clemson Univ., Pendleton, SC (dsahoo@clemson.edu); Research Assistant Professor, Clemson Univ., Georgetown, SC (lohallo@clemson.edu); Assistant Professor, Clemson Univ., Blackville, SC (lseveri@clemson.edu).

CHEMICAL DORMANCY AND COOL-SEASON ANNUAL FORAGE ESTABLISHMENT METHOD EFFECTS ON OVERSEEDED BAHIAGRASS

P. Zessin

Warm-season perennial pastures are the base of livestock production systems in lower Alabama. To overcome forage production shortages during bahiagrass (*Paspalum notatum*) dormancy in late fall and winter, overseeding pastures with cool-season annuals can provide forage. Diverse mixtures of species with effective establishment could increase production, but few evaluations have been conducted in overseeded sods. Our objective was to evaluate cool-season annual production using different overseeding methods in bahiagrass sod treated with different herbicides to induce dormancy. Herbicide treatments were two rates of paraquat and glyphosate and a control treatment (no herbicide) replicated in two planting methods (no-till drill or broadcast) of a mixture of annual ryegrass (*Lolium multiflorum)*, triticale (*Triticosecale*), and crimson clover (*Trifolium incarnatum*). There was no effect of treatment (planting method, herbicide, or herbicide rate) on forage biomass. Cool-season botanical composition was not affected by herbicide treatment or planting method. However, there was an effect of treatment in the warm season harvest (bahiagrass) in which the glyphosate treatment suppressed bahiagrass production. Also herbicide treatment affected warm-season botanical composition in which glyphosate treated plots experienced increased weed presence. There was no effect of herbicide treatment on forage CP, NDF, or ADF for the cool-season harvests. However, there was an effect of harvest for all cool-season annual forage nutritive value parameters. These results indicate that consideration may need to be given to establishment methods using nonselective herbicides for chemical dormancy. However, this study demonstrates that overseeding bahiagrass sods with cool-season annuals can be an effective cool-season forage system for southeastern producers.

MICROBIAL INOCULANT EFFECTS ON WARM-SEASON PERENNIAL GRASS SILAGE

Joao Vendramini and Philipe Moriel[1]

Silage has been the preferred method of forage conservation in many regions of the world. Corn and sorghum are the main forages conserved as silage, but livestock producers in tropical and subtropical regions have relied on warm-season perennial grasses for conserved forage. However, these grasses have limited dry matter concentration, and morphological and chemical characteristics that reduce fermentation and conservation. Microbial inoculants have been developed that improve fermentation of corn silage but are recommended for grass silage with limited evidence for their efficacy. Research conducted in Florida tested seven commercial inoculants on Jiggs bermudagrass [*Cynodon dactylon* (L.) Pers] and although there were some treatment differences, there was minimum improvement in fermentation characteristics when treated silages were compared to the control. It is important to mention that the forage had relatively low DM (25%) and water-soluble carbohydrate concentrations (5%), which may have led to limited conditions for appropriate activity of the microbial inoculant and desirable fermentation. Conversely, a recent study testing the effects of the microbial inoculant SiloSolve FC (*Lactobacillus budchneri* LB1819 and *Lactococcus lactis* O224) on limpograss [*Hemarthria altissima* (Poir.) Stapf & CE Hubbard] silage reported that treated silage had lower abundance of *Clostridium* and *Pediococcus* and greater *Lactococcus* and *Bacillus* than the untreated control. Changes in microbial abundance resulted in inoculated silage with greater acetic and propionic acid and lesser butyric acid and NH_3-N concentrations, and greater in vitro DM digestibility and CP concentrations. Tailored recommendations for microbial inoculant species and strains may be necessary to increase the predictability of warm-season grass silage inoculation.

[1]Range Cattle Research and Education Center, University of Florida, Ona, FL.

ARKANSAS HAY VERIFICATION PROGRAM

B. Finch, J. Mitchell, B. Mills, K. Lawson, and R. Loy[1]

Hay production is a vital intersection between the animal and agronomic production sectors of the agricultural industry. Hay production requires agronomic management of forage that will be harvested and exported to supplement the forage requirements of livestock and equine during periods of limited pasture forage availability. The cost to maintain adequate nutrient levels to maintain high-yielding and quality hay can become overwhelming with variable fertilizer prices and the lack of a uniform price market. The Arkansas Hay Verification Program aims to showcase the potential positive return on investment of fertilizers following scientifically based University recommendations. This program is a collaborative effort between University Extension and Research personnel and hay producers to implement and verify the economical return of management recommendations provided by the University of Arkansas System Division of Agriculture (UADA). Over the past two years of the program, 6 of 8 fields in 2022 and 9 of 11 fields in 2023 had positive net returns of production using a uniform price and actual cost of production. The average net returns were $200.79/acre in 2022 and $75.83/acre in 2023 with a range from -$136.90 to $783.27 per acre. While there is a large variation in net returns due to inputs and environments the overall positive net return in both years exhibits the ability of UADA recommendations to maintain economical production of hay.

[1]Soil Fertility Specialist, Univ. of Arkansas System Div. of Agric., Cooperative Ext. Service, Little Rock, AR (bfinch@uada.edu); Livestock Marketing and Management, Univ. of Arkansas System Div. of Agric., Cooperative Ext. Service, Fayetteville, AR (jlmitche@uark.edu); Asst. Professor Ag. Economics and Agribusiness, Mississippi State Univ., Starkville, MS (b.mills@msstate.edu); Faulkner Co. Agent, Univ. of Arkansas System Div. of Agric., Cooperative Ext. Service, Conway, AR (klawson@uada,edu); Asst. Professor Ag. Economics and Agribusiness, Univ. of Arkansas System Div. of Agric., Cooperative Ext. Service, Little Rock, AR (rloy@uada.edu).

A DEMOSTRATION OF THE ESTABLISHMENT AND GRAZING MANAGEMENT OF SPRING OATS AT DOUBLE OO RANCH

J.D. Huff and S.F. Tabler[1]

Cool season annual forages are an important part of Arkansas forage production. These forages provide opportunities for Arkansas livestock producers to extend the grazing season. Spring oats have been recently studied by the University of Arkansas Division of Agriculture for opportunities to extend fall grazing as well as silage. This presentation will share an experience of the establishment and grazing management of spring oats on a cow/calf operation in western Arkansas. A 40-acre pasture on Double OO Ranch was identified as the site for the demonstration. The pasture has a warm season base of bermudagrass, crabgrass, and dallisgrass. Tall fescue had been encroaching into this pasture and is viewed undesirable by the ranch. Annual ryegrass had been planted on October 2023, but the stand was marginal and viewed unsuccessful. A chemical burndown with glyphosate (2.5 quarts/acre) was applied in early February 2024 to eliminate the existing competition. A prescribed fire was applied in early March 2024 to remove excessive thatch. Spring oats were planted immediately after the fire at a rate of 100 lbs./acre. Jerry oat was selected as the variety for the demonstration. An application of 80 lbs. of nitrogen per acre was applied immediately after planting. The spring oat provided cattle with high quality forage and multiple grazing sequences between late April and late May.

[1]NRCS State Grazing Lands Specialist, USDA-NRCS (AR) (jeremy.huff@usda.gov); Ranch Manager, Double OO Ranch (sam.tabler@gmail.com).

IMPLEMENTATION OF CLIMATE-SMART STRATEGIES TO SUPPORT SUSTAINABLE LIVESTOCK PRODUCTION IN SOUTH CAROLINA OPERATIONS

C. Garcia, L. Silva, L. Vieira, J. Adkison, and P. Agudelo[1]

In the Southeast, livestock systems rely on perennial forage systems requiring inorganic nitrogen fertilizer to sustain production and diet quality. This practice increases the livestock industry's carbon footprint and risks degradation under poor management. This on-farm study evaluated climate-smart practices (incorporation of clover, poultry litter application, and rotational grazing) to enhance production, feasibility, and sustainability of forage-livestock systems in South Carolina. This Clemson University-South Carolina State's pilot project assessed forage mass and greenhouse gas (GHG) concentrations (CH_4, N_2O, CO_2) at two on-farm sites. Two treatments [grass-N fertilizer (GN) and grass-legume (GL)] were imposed from March to July 2024. On-farm #1, GL had 6 and 33% lower CO_2 and N_2O emissions, respectively, and reduced CH_4 eightfold compared to GN treatment. On-farm #2, GL sequestered seven times more CH_4 but had 10% higher CO_2 and 2.6 times greater ($P=0.04$) N_2O emissions. Higher soil humidity and nitrogen fixation likely drove increased emissions. Results emphasize the importance of tailored approaches and accurate quantification to optimize climate-smart practices' environmental benefits.

[1]Postdoctoral Fellow, Clemson University, Blackville, SC (cvelaga@clemson.edu); Assistant Professor, Clemson University, Blackville, SC (lseveri@clemson.edu), Postdoctoral Fellow, Clemson University, Blackville, SC (lvieira@clemson.edu), Field Technician, Clemson University, Blackville, SC (dadkison@clemson.edu), Professor, Clemson University, Clemson, SC (pagudelo@clemson.edu).

ASSESSING INOCULATION STRATEGIES FOR COOL-SEASON ANNUAL LEGUMES

D. Furukawa, R. Balster, M. Wallau, N. Caram and C.L. Mackowiak[1]

Nodulation in forage legumes relies on available and compatible rhizobia bacteria in the rhizosphere. To ensure adequate rhizobia availability, seed companies developed pre-inoculated seeds; typically clay-based products with claims of extended shelf life. However, limited nodulation success has been observed in the field. This study aimed to address potential nodulation effectiveness of clay-based pre-inoculated vs peat-based inoculation in clovers. Pre-inoculated seeds from four clover species (Arrowleaf, *Trifolium veisculosum*; Berseem, *T. alexandrinum*; Red, *T. pratense*; and White, *T. repens*), were purchased from commercial sources in fall 2022 and 2023, and subjected to four inoculation treatments: clay-based pre-inoculated (T1), sterilized seeds (T2), sterilized and reinoculated with peated inoculant (T3), and sterilized + N fertilizer (T4). Seeds were grown in a greenhouse in pots containing autoclaved sand. Plants were evaluated for nodule number, below-ground (root and nodules) and above-ground biomass at 63 days-after-planting (DAP). Plants were fertilized with N-free nutritive solution, except T4 (nutritive solution contained N). Treatments T2, T3, and T4 had some nodulation, indicating the sterilization method was not entirely effective. However, nodulation probability was greater for T3 and T4 (60-70%) than for T1 and T2 (<40%). Nodulation probability was also lower in Arrowleaf than Berseem. Reinoculation (T3) and N fertilization (T4) increased nodulation and biomass for all species except Arrowleaf. Overall, reinoculation was more effective than relying on clay-based pre-inoculation.

[1]Grad. student, Soil, Water, and Ecosystem Sciences UF, Quincy, FL (daichi.furukawa@ufl.edu); Agronomy M.S., UF (rachel.balster@ufl.edu); Forage Specialist, Agronomy, UF, Gainesville, FL (mwallau@ufl.edu); Postdoc, Agronomy, UF, Gainesville, FL (ncaramfernandezv@ufl.edu); Soil Specialist, Soil, Water, Ecosystem Sciences, UF, Quincy, FL (echo13@ufl.edu).

USING A DRONE FOR PASTURE FROST SEEDING

J.M. Hartschuh[1], A.D. Thomas[2], C. Gelley[3], G.R. Ruff[4], and B.J. Campbell[5]

Frost seeding of grasses and legumes allows for pasture remediation without the need for tillage to improve forage quality and livestock carrying capacity. Traditionally frost seeding is done using a ground-based spreader mounted on a tractor or all-terrain vehicle. This can create challenges when the terrain is unsafe for the operation of manned vehicles due to the slope of the hills or wet soils that do not allow for the operation of these vehicles. Drone seeding may be an option to alleviate these challenges keeping workers safe and allowing for frost seeding in adverse weather conditions. This study was conducted at The Ohio State University Research Stations in Caldwell and Jackson, Ohio. Two seed mixes were used a clover mix that contained Red clover and Ladino clover along with a grass mix of Annual ryegrass and Orchard grass. Plots were 30 feet wide and over 200 feet long. The center 100 feet was evaluated for forage establishment. The seeding pattern of the drone was evaluated through pan-testing, the grass seed pattern was less uniform with a seeding rate range of 2.76-14.21 lbs/ac with an average of 6.09 lbs/ac across the spread pattern. The clover pattern was more uniform with a range of 5.21-9.79 lbs/ac with an average of 7.80 lbs/ac across the spread pattern. Drone seeding may require a separate pass for grass versus legumes but if weather conditions are appropriate can be as successful as ground-based frost seeding.

[1]Extension Field Specialist Dairy Management and Precision Livestock, The Ohio State University Extension, Fremont, OH (hartschuh.11@osu.edu); [2]Food Agriculture and Biological Engineering Graduate Student, The Ohio State University, Columbus, OH (thomas.5083@osu.edu); [3]Extension Educator Noble County, The Ohio State University Extension, Caldwell, OH (gelley.2@osu.edu); [4]Extension Field Specialist Beef Cattle and Livestock Marketing, The Ohio State University Extension, Caldwell, OH (ruff.72@osu.edu); [5]Assistant Professor, State Small Ruminate Extension Specialist, The Ohio State University Department of Animal Science, Wooster, Ohio.

CRABGRASS AND BERSEEM CLOVER AS WEED SUPPRESSANTS IN SORGHUM x SUDANGRASS PASTURES

Justin C. Burt[1], Kathy J. Soder[2], and Kelly M. Mercier[1,3]

Sorghum-sudangrass [*Sorghum bicolor* × *S. bicolor var. sudanense*; SxS] is a common warm-season annual forage grown in the northeastern US for grazing. Utilization of other high-quality annual grasses and legumes, such as crabgrass (*Digitaria sanguinalis* L.; CG) and berseem clover (*Trifolium alexandrinum* L.; BC), are of interest to aid in weed control and potentially improve the production of SxS forage systems. A two-year (2022 and 2023) small-plot evaluation was conducted in University Park, PA, to determine the agronomic benefits of interseeding CG and BC into SxS established on different row spacings. Plots were established in a randomized complete block design with three replications. Factors evaluated consisted of 1) SxS planted on a 7.5- or 15-in. row spacing and 2) SxS monoculture or SxS interseeded with CG, BC, or CG + BC, that had been either 3) drilled or broadcast planted into the SxS, resulting in eleven treatment combinations. Seasonal herbage accumulation did not differ among treatments for either year of the study ($p > 0.65$). Botanical composition differed among treatments ($p < 0.01$). In plots interseeded with CG there was a lower percentage of SxS (~45%), but also lower instance of weeds in the stand (~10% to 15%). Berseem clover consisted of less than ~10 % of the mixtures and did not reduce weed competition. Future research should focus on other legume species interseeded into SxS, since BC was not effective at weed suppression, and evaluate these mixtures under different defoliation methods, such as grazing.

[1]Postdoctoral Researcher (justin.burt@usda.gov) and [2]Research Animal Scientist, (kathy.soder@usda.gov), USDA-ARS Pasture Systems and Watershed Management Research Unit, University Park, PA; [3]Current address: Grazing Specialist (kelly.mercier@usda.gov), USDA-NRCS, Bloomsburg, PA.

IMPACT OF RED-CLOVER ROW SPACING IN BERMUDAGRASS

Justin C. Burt[1] and Lisa L. Baxter[2]

Interseeding clovers (*Trifolium* spp.) into bermudagrass (*Cynodon* spp.) is not a novel concept. The development of a biennial red clover (*Trifolium pratense*) has producers interested in a lower cost legume in comparison to alfalfa (*Medicago sativa* L.) for their pastures. However, data are limited on the optimal row spacing of this type of clover in bermudagrass based forage systems. Therefore, the objective of this study was to compare the agronomic benefits of red clover interseeded into bermudagrass. A small plot study was conducted in a randomized complete block design with four replications at the University of Georgia-Tifton Campus, Tifton, GA. Treatments consisted of bermudagrass fertilized with N, and bermudagrass interseeded with 'Barduro' red clover on a 14-, 21-, or 28-in row spacing. Data indicate that the seasonal herbage accumulation was impacted by the treatment ($p < 0.01$). Seasonal herbage accumulation was greater in bermudagrass plots receiving N fertilizer, compared to other treatments. Bermudagrass plots interseeded with red clover despite different row spacing reported a similar herbage accumulation to each other ($p = 0.05$). Nutritive value parameters of crude protein and total digestible nutrients, however, did not differ among treatments ($p > 0.39$). The results of this first year show promise and the study will be repeated for a second year. Once the optimal row spacing is determined, future research is warranted to determine the optimal harvest height and frequency for this mixture.

[1]Postdoctoral Researcher, Crop and Soil Science Department (justin.burt@uga.edu); University of Georgia Tifton Campus, Tifton, GA. [2]Assistant Professor and Extension Forage Specialist, Crop and Soil Science Department (baxterl@uga.edu); University of Georgia Tifton Campus, Tifton, GA.

LATE WINTER COOL-SEASON ANNUAL FORAGE PLANTINGS FOR SPRING PRODUCTION

K.J. Simon, D. Kennedy, and J.O.C. Kubesch[1]

Cool-season annual grasses are often overseeded into bermudagrass pastures in the fall or late winter to extend the growing season in the southeastern USA. Late winter plantings can provide an early hay crop or forage for grazing where perennial stands have been affected by adverse weather conditions. The optimum late winter planting date is not well established for these annual forages. Our objective was to compare three late-winter planting dates on forage mass of four annual grasses. This experiment was conducted at the Arkansas State University farm in Jonesboro, AR from 2021-2024. Winter wheat, spring and winter oats, and annual ryegrass were no-till planted into a glyphosate-suppressed bermudagrass sod on one of three planting dates. Early planting dates were in late February, Mid planting dates were in mid-March, and Late planting dates were in late March. Plots were harvested once in late May of each year. Winter wheat was the least productive forage across planting dates in all years. Spring oats were the most productive forage. The early and mid-planting dates produced similar amounts of forage at time of harvest. Late plantings were less productive. These results were consistent despite the variability in weather conditions across multiple seasons. This experiment suggests that the optimal planting period for late winter-planted annual forages is between late February and mid-March in the southeastern USA.

[1]Forages Instructor, University of Arkansas System Division of Agriculture, Cooperative Extension Service, Little Rock, AR (ksimon@uada.edu); Forage Specialist, University of Arkansas System Division of Agriculture, Cooperative Extension Service, Little Rock, AR (jkubesch@uada.edu); Associate Dean and Professor of Animal Science, A-State University, Jonesboro, AR (dkennedy@astate.edu).

ESTABLISHING ALFALFA WITH A SUMMER ANNUAL NURSE CROP

Lauren Jaschke and Shelby M. Gruss[1]

Alfalfa (Medicago sativa) is an important forage crop, providing a high quality, high yielding, perennial forage for livestock producers across the U.S. Nurse crops can aid alfalfa establishment by providing ecosystem services such as reducing nutrient runoff and controlling soil erosion while potentially increasing production during establishment year. Summer annual grasses, known for their high yield, forage quality, and rapid growth, could serve as valuable nurse crops in a multi-cut system by enhancing light penetration to facilitate alfalfa establishment. This study investigates the use of summer annuals as nurse crops for fall- and spring-planted alfalfa. Two summer annuals, sorghum x sudangrass (SxS), and sudangrass, were evaluated as a nurse crop for three alfalfa varieties with two alfalfa planting dates; spring, and late summer. As controls, summer annuals and alfalfa monocultures were established. Spring planted alfalfa was planted with the summer annuals (06/14/2024), while the late summer was planted following the first summer annual harvest (08/06/2024). Summer annuals were harvested three times, with biomass collected at each harvest. At the final harvest alfalfa biomass was collected from alfalfa monocultures. Lastly, alfalfa stand counts were collected in the fall for all treatments. SxS had a higher average yield per harvest compared to sudangrass, 1.57 tons acre-1 and 1.37 tons acre-1 respectively (P-value <0.05). Alfalfa monocultures had higher stand counts in comparison to the intercropped plots, suggesting competition. Spring 2025 alfalfa harvest will elaborate the impacts of the summer annual competition on the alfalfa's future productivity.

[1]Department of Agronomy, Iowa State University, Ames, IA 50011, USA (ljaschke@iastate.edu, sgruss@iastate.edu).

EVALUATION OF *BRACHIARIA (UROCHLOA)* CULTIVARS AS AN ANNUAL SYSTEM IN THE SOUTHERN USA

R. Lemus[1] and J. White[2]

Forage productivity is diminished under warm and humid conditions, affecting growth, fertilizer use efficiency, and nutritive value. We determined the response of *Brachiaria (Urochola)* cultivars to nitrogen fertilization as an annual forage system. Nitrogen (N) was applied at 0, 60, 90, and 120 lb N ac^{-1} in split applications as urea ammonium sulfate (33-0-0-18S). The cultivars were 'Basilisk' (*U. decumbens*), 'BRS Piata' (*U.* brizantha), Camello Blend' (*U.* ruziziensis × U. decumbens × U. brizantha), 'Cayana' (*U.* hybrid), 'Cobra' (*U.* hybrid cv. CIAT BR02/1794), 'Convert 330' (*U. brizantha x U. ruziziensis*), "Marandu' (*U. brizantha*), 'Mestizo Blend' (*U.* hybrid cv. CIAT 36087/GP3025/GP3207) 'Mulato II' (*U. ruziziensis x U. brizantha x U. decumbens*). *'Ruziziensis' (U.* ruziziensis), and Toledo (*Brachiaria brizantha*). Cultivars were seeded June 28, 2023 in a prepared seed bed at 10 lbs PLS/ac. Green canopy cover (GCC, %), leaf area index (LAI), and normalized difference vegetation index (NDVI) were measured. Plots were harvested in August and September. Nutritive analysis was CP, ADF, NDF, IVTDMD, WSC, and starch. Seasonal biomass production was highest for Cobra and Basilisk. Biomass production peaked at 60 lb N/ac with a 50% nitrogen use efficiency and a leaf area index of 5.8. There were cultivar differences in nutritive value ranging from 15.9 to 19.1% CP, 28.6 to 30.5% ADF, 51.1 to 8.4% NDF, 78.6 to 85.5 IVTDMD, and 5.9 to 7.4 % WSC, and with N rate peaking at 60 lb N ac^{-1}. Preliminary data suggests that Brachiaria cultivars might have an opportunity to be used in the southern USA as a late-season pinpoint forage option.

[1]Professor and Extension Forage Specialist, Mississippi State University, MS (Rocky.Lemus@msstate.edu); [2]Forage Variety Testing Manager, Mississippi State University, MS (JWhite@pss.msstate.edu).

SMOTHER CROP EFFICACY OF SORGHUM-SUDANGRASS AND PEARL MILLET WITH AND WITHOUT THE BROWN MID-RIB TRAIT

S. Dickerson, K. J. Simon, J.O.C. Kubesch, and M.S. Gadberry[1]

Warm-season annual grasses are the primary smother crops used in pasture renovation. Smother crops often serve dual roles as competitors against weeds but also as livestock forage. This experiment compared botanical composition and weed suppression of sorghum-sudangrass and pearl millet, with and without the BMR trait, under two different harvest management regimes at the Livestock and Forestry Research Station in Batesville, AR. Plots were no-till planted on 5/30/24 at 25 lb./acre and managed as either common producer practices (CPP) or best management practices (BMP). Stand composition was assessed using frequency grids. Plots with the BMR trait initially had higher frequencies of crop plants than plots without BMR. At the time of final harvest, BMR stands had higher crop frequencies under CPP than under BMP. Sorghum-sudangrass stands had higher crop frequencies than pearl millet stands at the time of final harvest. The stand composition was similar between the first and second harvests. The overall stand composition changed at the time of the third harvest, which reflects the seasonal emergence of warm-season grassy weeds and cool-season broadleaf weeds in the plots. These results suggest that harvest management can determine smother crop efficacy. There does not appear to be an advantage to selecting BMR pearl millet or sorghum-sudangrass varieties over standard options for the purpose of a smother crop.

[1]Jackson County Agent, University of Arkansas System Division of Agriculture, Cooperative Extension Service, Forrest City, AR (sdickerson@uada.edu); Forages Instructor, University of Arkansas System Division of Agriculture, Cooperative Extension Service, Little Rock, AR (ksimon@uada.edu). Forage Specialist, University of Arkansas System Division of Agriculture, Cooperative Extension Service, Little Rock, AR (jkubesch@uada.edu); Director of the Livestock and Forestry Research Station, University of Arkansas System Division of Agriculture, Batesville, AR (sgadberry@uada.edu).

ERGOT-INFESTED SEEDHEADS IN FLORIDA BAHIAGRASS PASTURES

Benjamin Reimer, Haihua Wang, Kaile Zhang, Vijay Verma, Valerie Mendez, Ann Blount, Brittany Justesen, Joeseph Walter, Cheryl Mackowiak, Robbie Jones, Anna Paulette Tomlinson, James T DeValerio, Doug Mayo, and Hui-ling Liao

Ergot is a common fungal disease found in Florida pastures of Argentine bahiagrass that may become a greater threat for future cattle farmers. Cattle ingestion of infected seed heads may lead to reproduction issues, weight loss, reduced milk production, and neurological symptoms of staggered movements. While there is an extensive body of knowledge on the disease and appropriate treatment responses, we are concerned for new varieties and shifting seasonal distributions of the fungi *Claviceps* which causes the ergot disease. Here we review and introduce the current knowledge necessary for cattle managers to recognize and respond to outbreaks of the disease in their pastures. We additionally provide our own findings for epidemiological distribution and morphology of the disease in Florida, as well as new details on *Claviceps* varieties that we've identified in North Florida. We detected four ergot alkaloids (ergotamine, ergonovine, elymoclavinein, and dihydrolysergol) in our bahiagrass leaf samples. The concentrations of these alkaloids ranged from 1 to 763 ppb, affecting between 1.32% and 6.58% of the samples. Our findings enhanced the understanding of the pathogenic mechanisms of ergot disease on bahiagrass seedheads and its threat to cattles.

EFFECT OF APPLICATION OF PROPIONIC ACID FORAGE PRESERVATIVE AT MOWING IN ALFALFA-BERMUDAGRASS MIXTURES PRESERVED AS BALEAGE ON THE MIXED RUMINAL MICROORGANISMS FERMENTATION IN VITRO

B.A. Stefancik[1], T.R. Callaway[2], and J.J. Tucker[1]

Southeastern producers are interested in utilizing forage preservatives; however, to date, ruminal degradation of preservative treated forages has not been widely evaluated. An in vitro ruminal fermentation study was conducted in April 2024 utilizing forage preserved in individual mini-silos from a July 2022 cutting of 'Bulldog 805' alfalfa interseeded into 'Tifton-85' bermudagrass (ABG). The objectives were to evaluate in-vitro ruminal fermentation for 1) ABG sampled at different time points post-harvest and 2) ABG receiving forage preservative application. The treatments evaluated were ABG receiving propionic acid forage preservative (P+) or no preservative (NP) at mowing and then sampled at harvest (H), 8 weeks post-harvest (8W), and 6 months post-harvest (6M). Forage was lyophilized and ground to pass a 1 mm screen on a Wiley grinder. Samples were fermented in individual anaerobic mixed ruminal microorganism fermentations in bottles sealed with butyl rubber stoppers and aluminum crimps in triplicate at 4 time points: 2, 4, 24, and 48 hours of fermentation. Total gas production was higher for H (46 mL) and P+ (47 mL) as compared to 6M (43 mL) and NP (43 mL), respectively, after 24 hours of fermentation ($p < 0.04$). Methane production was higher for P+ (7.15 mmol) as compared to NP (6.61 mmol; $p < 0.01$). IVDMD was higher for NP forage (68.5 %) as compared to P forage (66.7%, $p < 0.01$). This preliminary data suggests that application of preservatives to forages does not improve ruminal fermentation; however, future work looking at feeding cattle ABG baleage is warranted to expand upon this introductory experiment.

[1]Animal and Dairy Science Department, University of Georgia, Tifton, GA; 2 Animal and Dairy Science Department, University of Georgia, Athens, GA.

ENDOPHYTES AND MYCOTOXINS IN FLORIDA PASTURES

K. Zhang[1], H-L Liao[1], H. Wang[1], B. Reimer[1], V. Verma[1], B. Justesen[2], J. Walter[2], A. P. Tomlinson[2], R. Jones[2], V. Mendez[2], A. Blount[1], C. Mackowiak[1], M. Wallau[2], K.-H. Chen[1], J. Duringer[3], J. Love[1], J.K. Yarborough[2], and D. Mayo[2]

Grass leaves host diverse fungal communities, especially endophytes, which can produce secondary metabolites to enhance grass health and survival under biotic and abiotic stress. However, some secondary metabolites may transform into harmful mycotoxins when ingested by livestock at high concentrations. The influence of various abiotic and biotic factors on the dynamics of grass leaf endophytes and mycotoxins remains largely unexplored. To address this gap, we conducted a large-scale study across 13 ranches in Florida over different seasons from 2017 to 2019 to characterize variations in leaf fungal communities and mycotoxins among four pasture grass species (bahiagrass, bermudagrass, limpograss, and smutgrass). We observed distinct biogeographic and climatic patterns in the distributions of leaf fungal communities and mycotoxins, along with their interactions. Notably, zearalenone-4-sulfate, ochratoxin α, and emodin were consistently found at high concentrations across all sampled sites, species, and seasons. Furthermore, grass species significantly influenced the distribution of mycotoxins, with bahiagrass exhibiting relatively lower concentrations compared to other species.

[1]North Florida Research and Education Center, University of Florida, 155 Research Road, Quincy, FL 32351, USA (kaile.zhang@ufl.edu, sunny.liao@ufl.edu, wanghaihua@ufl.edu, reimerb@ufl.edu, v.verma@ufl.edu, paspalum@ufl.edu, echo13@ufl.edu, kohsuanchen803@gmail.com); [2]Institute of Food and Agricultural Sciences Extension, University of Florida, Gainesville, FL 32611 (brittanyjustesen@ufl.edu, jwalter@ufl.edu, r.jones1@ufl.edu, apt@ufl.edu, valeriemendez@ufl.edu, mwallau@ufl.edu, jclove@ufl.edu, jyarborough@ufl.edu, demayo@ufl.edu); [3]Department of Environmental and Molecular Toxicology, Oregon State University, Corvallis, OR 97331, USA (jennifer.duringer@oregonstate.edu).

ANNUAL BEEF AND FORAGE FIELD NIGHT EVALUATION FOR CLIENTELE

J. Penrose, J. Winters, D. Dempsey, C. Bruynis, T. Corboy, G. Ruff, T. Wiseman, J. Hartschuh, and C. Penrose[1]

Extension Educators in this region identified that beef and forage production is a primary focus and have partnered with the Ohio Agricultural Research and Development Center in Jackson, Ohio. Educators provide educational programming related to beef and forage production to improve farm profitability and sustainability in southeastern Ohio. A quantitative approach was taken where participants were asked to rank learning statements on a 1 – 6 Likert scale on a retrospective pre/post-test. Learning statements included: (1) I understand the opportunities and challenges of using drones for pasture management. (2) I understand additional management systems that can be added when using EID tags on cattle. (3) I understand the economic benefits of implementing grazing strategies. (4) I know the environmental benefits that grazing can provide. (5) I know how to use data in a sire directory. (6) I know how to identify economically important traits specific to a herd. (7) I understand proper gate placement in my fence. (8) I understand how to select the correct size fence energizer. (9) I understand what type of staple I have and correct placement. (10) I understand the importance of proper corner post installation. A net gain of knowledge of 1.74, improving on the 2022 results of 1.66.

[1]Ohio State University Extension Educator, McConnelsville, OH (penrose.30@osu.edu); Ohio State University Extension Educator, Jackson, OH (winters.249@osu.edu); Ohio State University Extension Educator, Piketon, OH (dempsey.106@osu.edu); Ohio State University Extension Educator, Chillicothe, OH (bruynis.1@osu.edu); Ohio State University Extension Educator, Georgetown, OH (corboy.3@osu.edu); Ohio State University Extension Field Specialist, Caldwell, Ohio (ruff.72@osu.edu); Ohio State University Extension Educator, New Lexington, OH (wiseman.15@osu.edu); Ohio State University Extension Field Specialist, Fremont, OH (hartschuh.11@osu.edu); Ohio State University Extension Emeritus, Stockport, OH (penrose.1@osu.edu).

FORAGES IN THE LAST FRONTIER: A SURVEY OF FORAGE NUTRITIVE VALUE IN ALASKA

Kacey Aukema and Rachael Christensen[1]

Forages make up the largest component of ruminant livestock diets. Therefore, the nutritive value of forages has a large impact on growth, development, and productivity of livestock. In the extreme climate of Alaska, the growing season for pasture and hay crops is very short. Subsequently, forage resources are a very limited and valuable commodity. Given the geographic isolation of the livestock and hay production industry in the state, there is relatively limited livestock nutrition guidance support available locally. A survey of forage nutritive value of pasture and hay crops in the agricultural communities of Alaska was conducted to better understand the nutritive value status and trends of forage resources of the state. In turn, this could potentially guide supplementation strategies to use the limited forage resources of the area more efficiently. Variation in forage nutritive value trends over time during the season were influenced by principal forage species present (Bluegrass or Brome). Current nutritive value survey results compared to historic values presented in literature from previous surveys show modest improvement in some parameters.

[1]Agricultural Science Technician (kacey.aukema@usda.gov); Research Animal Scientist (rachael.christensen2@usda.gov); USDA-ARS, Northern Great Plains Research Laboratory, Mandan, ND. Special thanks to Phillip Kaspari, Eve Karczmarczyk, Milan Shipka, and Raina Hanley and the various forage and livestock producers of Alaska that allowed to us to sample their hay and pastures.

CASE STUDY: MILK UREA NITROGEN ON ORGANIC GRASSFED DAIRY FARMS

Kathy Soder[1], Kelly Mercier[1,2], Sara Ziegler[3], Sarah Flack[4], and Heather Darby[5]

The ratio of energy to protein in the diets of grazing dairy cows may not be balanced for optimal milk production and is generally monitored through milk urea nitrogen (MUN) concentration. Grassfed dairies, where grain is not fed, often have highly variable MUN values that are outside the ideal range (10-14 mg/dL) which may impact animal productivity and health. A case study was conducted on 5 organic grassfed dairies in the Northeast over 2 grazing seasons to identify animal, feed, or management factors that may be influencing MUN. Correlations were conducted between MUN and milk production variables, reproductive traits, and body condition. Milk yield was positively correlated with MUN on four farms, with a negative correlation on one farm. Milk fat yield was positively correlated with MUN on four of the farms, while milk protein yield was positively correlated with MUN on only 2 of those farms. Four of the 5 farms had a positive correlation between MUN and energy-corrected milk. Reproductive traits, at best, had a weak negative correlation with MUN. Pasture crude protein levels (CP) exceeded (19% CP average across farms with a range of 13-24% CP) recommended levels for lactating dairy cows (15-17% CP) but were not consistently correlated with MUN. Based on these preliminary results, MUN is a complex issue that may be influenced by multiple animal and forage management factors that requires more in-depth study.

[1]Research Animal Scientist, USDA-ARS, Pasture Systems & Watershed Management Research Unit, University Park, PA (kathy.soder@usda.gov); [2]Former USDA-ARS post doc- Current address: Grazing Specialist, USDA-NRCS, Bloomsburg, PA (kelly.mercier@usda.gov); [3]Research Specialist, University of Vermont, Burlington (sara.ziegler@uvm.edu); [4]Sarah Flack Consulting, Fairfield, VT (sarahflackconsulting@gmail.com); [5]Agronomy Specialist, University of Vermont (heather.darby@uvm.edu).

AGRONOMIC RESPONSES AND ANIMAL PERFORMANCE OF BEEF REPLACEMENT HEIFERS GRAZING COOL-SEASON MIXTURES: A CASE STUDY

K. Seavey, J. Adkison, C. Garcia, and L. Silva[1]

In South Carolina, cool-season annual forages are crucial to extend forage production after perennial warm-season grasses become dormant. Forage mixtures are commonly used to provide better forage production and distribution, and the inclusion of legumes helps to reduce nitrogen (N) fertilizer input and improve animal performance. This case study aimed to determine forage mass, nutritive value, and animal performance under contrasting cool-season forage mixtures. Two mixtures were used: GRASS [rye (Secale cereale), triticale (x Triticosecale W.) and ryegrass (Lolium multiflorum)], and GLEG [rye, triticale, ryegrass, crimson clover (Trifolium incarnatum) and hairy vetch (Vicia villosa R.)]. Both herbage mass (HM) and botanical composition were determined on pre-grazing condition. Black Angus-Cross heifers were used and weighed every 14 days to estimate average daily gain (ADG). On grazing cycles 1 and 2, the HM (P=0.38 and P=0.12, respectively) for the GRASS and GLEG treatments averaged 3135 vs 2646 and 5887 vs 3719 lbs./ac-1, respectively. The ADG (P=0.29) for the GLEG and GRASS treatments averaged3.85 and 3.38 lbs./d, respectively. Although no herbage mass differences were observed, including legumes in mixtures allows residual N to be left for the next crop or season in rotation. Furthermore, an economic analysis is necessary to ascertain the feasibility of using grass-legume mixture for beef replacement heifer diets and their environmental contributions to the ecosystem.

[1]Master Student, Clemson University, Blackville, SC (kseavey@clemson.edu); Postdoctoral fellow, Clemson University, Field Technician, Blackville, SC (cvelaga@clemson.edu); Clemson University, Blackville, SC (dadkiso@clemson.edu), Assistant Professor, Clemson University, Blackville, SC (lseveri@clemson.edu).

PHYTOESTROGEN CONCENTRATIONS IN RED CLOVER RESPOND TO ELEVATED TEMPERATURE BUT NOT CO_2

Palash Mandal[1], Nicholas D. Warren[2], Marta R.M. Lima[3], Andre F. Brito[1], and R.G. Smith[2]

Legumes are a high quality forage; however, some species synthesize and accumulate phytoestrogens, secondary compounds that can impair animal reproductive performance when consumed in large quantities. Previous research indicates that phytoestrogens may play roles in mediating plant responses to stress, suggesting that phytoestrogen concentrations in phytoestrogenic forage legumes may be sensitive to stress factors associated with climate change. This study investigated the effects of elevated temperature (*e*T) and elevated CO_2 (*e*CO_2), alone and in combination, on phytoestrogen concentrations in two forage legume species, red clover (*Trifolium pratense L.*) and cowpea (*Vigna unguiculata* L. Walp). Plants were grown in growth chambers under ambient conditions and then subjected to *e*T (35/26 °C, Day/Night) and *e*CO_2 (750±50 ppm) for a period of ten days, after which time tissues were sampled to determine phytoestrogen concentrations. Results demonstrated that total phytoestrogen concentrations in red clover were decreased by 50% under *e*T, while concentrations of individual phytoestrogen compounds varied in their response to *e*CO_2. In contrast, phytoestrogen concentrations in cowpea tissues remained below the level of detection in all treatments. Results of this study suggest that short-term increases in atmospheric temperature (i.e., heatwaves) associated with climate change may lead to a reduction in total phytoestrogen concentrations in red clover, and cowpea utilized as forage may not be a significant source of phytoestrogen exposure.

[1]Department of Agriculture, Nutrition, and Food Systems, University of New Hampshire, Durham, NH; [2] Department of Natural Resources and the Environment, University of New Hampshire, Durham, NH; [3]School of Animal Sciences, Virginia Tech, Blacksburg, VA.

EFFECTS OF FORAGE TYPE AND IONOPHORE PROVISION ON GRAZING CATTLE

P.A. Gunter[1], A.M.P. Netthisinghe[1], and K.R. Sistani[2]

Forage quality differs between cool-season annual and perennial grasses, with annual forages generally containing greater concentrations of protein. Ionophores are a feed additive that alters rumen microbial populations can improve the growth efficiency of cattle through increased energy and protein metabolism. This study evaluated how feeding ionophores impacted cow weight gain and manure composition when grazing either tall fescue (*Festuca arundinacea*) or annual ryegrass (*Lolium multiflorum*) pastures. Mature angus crossed cows grazed pasture for 45 days. Ryegrass and tall fescue pastures were not different in terms of crude protein content (10.8 vs. 11.7%), but tall fescue had higher fiber concentrations compared with ryegrass (30.2 vs. 19.3% ADF; 56.5 vs. 37.6% NDF). Digestibility was greater for ryegrass than tall fescue grass (76.6 vs. 67.6% total in vitro digestibility). Total gain and average daily gain (ADG) were greater for ryegrass (164 lbs.; 3.64 lbs./d) than tall fescue (109 lbs.; 2.42 lbs./d). Interestingly, for both forages total gain and ADG were greater for cattle that did not receive monensin compared with those that received monensin. The difference for ryegrass was 43 lbs. total gain and 0.99 lbs. ADG, and 21 lbs. total gain and 0.46 lbs. ADG for tall fescue. This may result from the high quality of the forages grazed and the relatively short grazing period. Monensin inclusion did not affect manure percent N, P, or K for ryegrass (0.395, 0.135, 0.155) or tall fescue (0.390, 0.130, 0.130).

[1]Dept. of Agriculture and Food Science, Western Kentucky University, Bowling Green, KY (phillip.gunter@wku.edu); [2] USDA-ARS Food Animal Environmental Systems Research Unit, Bowling Green, KY.

PASTURE MANAGEMENT AND COPPER UTILIZATION OF ANGUS HEIFERS

R.G. Christensen, K.D. Aukema and J.R. Hendrickson[1]

Meeting nutrient recommendations, specifically mineral requirements, is important for young livestock to obtain desired growth. Forages in rangelands tend to be deficient in some trace mineral content and require additional supplementation. Therefore, a wide variety of supplements are available, but little is known which specific mineral concentrations and mineral sources are better for rangelands, and what mineral content forage on rangelands might provide during the grazing season. We evaluated two different sources of copper mineral supplements and two types of grazing management practices on mineral absorption by growing beef heifers. Twenty-four Angus heifers were randomly assigned to plots of predominantly cool season grass pasture (brome and bluegrass). Half of the plots were located on pastures grazed yearly, while the other half on pastures allowed to rest at least 1 year between grazing. Two different mineral sources, one containing copper chloride with the other containing copper amino acid complex were fed. Individual liver biopsies were taken before and after grazing to compare liver copper absorption. Results showed that management had little effect on forage mineral content, but mineral source affected liver mineral content and average daily gain. Liver mineral content tended to increase for heifers provided amino acid complexed copper source. Results also suggest that mineral is overfed but must be supplemented to meet mineral requirements. There are indications that antagonism of mineral is present in these pastures.

[1]Research Animal Scientist (rachael.christensen2@usda.gov); Agricultural Science Technician (kacey.aukema@usda.gov); Research Rangeland Scientist (john.henrickson@usda.gov), USDA-ARS, Northern Great Plains Research Laboratory, Mandan, ND.

EVALUATION OF MANAGEMENT STRATEGIES FOR WINTER FEEDING AND GRAZING ON BEEF COW-CALF PRODUCTIVITY

S.M. Justice[1], J.D. Rivera[2], R.C. Jones, C.T. Shelton[2], and G. Gourley[2]

This study investigates the impact of different winter management strategies on cattle and pasture performance in Southern Arkansas, aiming to improve the efficiency and sustainability of cattle operations through better grazing and nutritional management. The trial evaluated three winter management systems: rotationally grazed cool-season annuals planted by traditional methods (TCSA), rotationally grazed cool-season annuals planted by broadcasting seed (BCSA), and free-choice bermudagrass hay with supplementation (HSUP). The objectives were to assess the system effects on cow body condition scores (BCS), calf performance, and economic costs. Thirty multiparous winter-calving beef cows were randomly assigned to the three treatment groups. Grazing treatments (TCSA and BCSA) involved rotational grazing of a ryegrass-rye mixture, with cows rotating through paddocks and receiving free-choice bermudagrass hay. Stocking rates were adjusted based on forage availability, and grazing ceased when pasture canopy heights fell below 10 cm. The HSUP group had free-choice access to bermudagrass hay supplemented with additional feed (up to 1% of body weight per day). Animal performance was measured through BCS assessments at project initiation, and at the end of the grazing season, as well as calf weight measurements at birth and post-grazing. While cattle performance did not differ across treatments, cost differences were higher for the HSUP system. This study provides insights into the establishment and management of cool-season forages and cattle performance, refining current recommendations for winter feeding strategies in beef cattle operations.

[1]University of Arkansas System Division of Ag Extension, Little Rock, AR (mjustice@uada.edu); [2] University of Arkansas System Division of Ag SWREC, Hope, AR.

EFFECTS OF PEANUT SKIN SUPPLEMENTATION ON HEIFERS GRAZING WILDTYPE VS. NOVEL ENDOPHYTE FESCUE

S. Poudel, G.J. Pent, L. Wright, J. Fike, and W. Zeller[1]

Fescue toxicosis (FT), caused by wild-type endophyte-infected (WE) tall fescue, causes significant financial and environmental costs, highlighting the need for cost-effective mitigation strategies. Some studies suggest condensed tannins (CT) may help alleviate FT, and this study evaluated the potential of peanut skin (PS) pellets, a by-product rich in CT, to alleviate the effects of FT in beef cattle. In the summer of 2023, 40 Angus and Angus-cross heifers were assigned to either WE or novel endophyte-infected (NE) tall fescue pastures, with or without PS supplementation, for 12 weeks. We measured body weight, hair cortisol (COR) levels, and urinary concentrations of lysergic and iso-lysergic acids at four-week intervals. Heifers grazing WE without PS had lower ADG (0.5 kg/d, $p < 0.05$) than those on NE (0.7 kg/d), with or without PS supplementation. PS supplementation significantly reduced COR in WE-grazing heifers ($p < 0.05$) but did not affect urinary toxin levels, which were higher in the WE group compared to those grazing NE, regardless of supplementation ($p < 0.01$). These findings suggest that while PS supplementation can alleviate certain stress indicators associated with FT, it may not fully counteract the negative effects observed in animals that consume WE-infected fescue pastures.

[1]Extension Animal Production Specialist, North Carolina A&T State University, Greensboro, NC (spoudel1@ncat.edu); Superintendent, Shenandoah Valley Agricultural Research and Extension Center, Virginia Tech, Raphine, VA (gpent@vt.edu); Superintendent, Southwest Virginia Agricultural Research and Extension Center, Virginia Tech, Glade Spring, VA (lrite@vt.edu); Professor, School of Plant and Environmental Sciences, Virginia Tech, Blacksburg, VA (Jfike@vt.edu); US Dairy Forage Research Center, USDA-ARS, Madison, WI (wayne.zeller@usda.gov).

AN ASSESSMENT OF YIELD AND QUALITY OF STOCKPILED BAHIAGRASS IN CENTRAL ALABAMA

A.J. Hurst, M.K. Mullenix, and S.L. Dillard[1]

Stockpiling is the practice of deferring the use of a grazing pasture for a period to allow the growth of a selected forage until frost. This practice helps bridge the forage production gap in the fall or early winter until cool-season forages are established. The objective of this study was to evaluate forage production and nutritional value of stockpiled bahiagrass and to determine its potential to extend the grazing season by providing additional days of forage in the fall months. In this study, two farm locations were utilized: Montgomery, AL, and St. Clair, AL. While both producers followed standard practices for stockpiling, the producer in St. Clair, AL applied 60 lbs/ac of fertilizer, while the Montgomery producer did not. Forage samples were collected biweekly during the stockpiling period to determine forage production (lb DM/acre) and nutritive value (*in-vitro* digestibility, crude protein, neutral detergent fiber, acid detergent fiber, acid detergent lignin). Variables were analyzed using GLIMMIX feature in SAS 9.4. At the study locations, producers were able to provide grazing for 31 and 36 days, respectively. The producer in St. Clair, who fertilized their stockpiled bahiagrass, achieved on average approximately 887 lb DM/ac greater biomass production, which correlated to the greater forage heights. Conversely, the producer in Montgomery, who did not apply fertilizer, still maintained moderate nutritive value, (7.3167% crude protein and 62.56% TDN). Results from this study indicate that stockpiling bahiagrass demonstrated potential to help close the fall forage gap, however protein supplementation may need to be provided.

[1]Department of Animal Science, Auburn University, Auburn, AL.

WARM-SEASON ANNUALS FOR SHEEP IN D4 DROUGHT

C.H. Gelley[1]

Warm-season annual forages are frequently recommended to extend the grazing season through the warmest months of the year in Ohio. While many options are available to plant, there is little quantifiable data available to demonstrate successful grazing with sheep. 2024 presented an opportunity to experiment with sheep grazing warm-season annuals amidst the most significant drought period that southeastern Ohio has experienced in 100 years. Four treatments were implemented: sorghum-sudangrass, pearl millet, a warm-season summer grazing mix, and established tall fescue, which was the control. The warm-season annual paddocks were seeded on July 25, 2024 and ready to graze on September 4, 2024, despite only receiving 2.2 inches of rain in 41 days. Based on the dry matter available of each forage type and a projected 7-day allocation of forage, the experimental area was projected to provide enough forage to strip graze each treatment over the course of 4 weeks. Stocking density was matched to the dry matter available. Mature ewes were turned in to graze on September 10, 2024 and by September 30, 2024 grazing was complete. Animal weight and body condition score was recorded before grazing and after grazing. Forage mass was recorded before and after grazing each allocation. Forage quality was also documented. Grazing animal days were calculated by multiplying the number of animals maintained on pasture by the number of days each treatment was grazed. Results by treatment ranged from 120 grazing animal days on pearl millet to 240 grazing animal days on the annual mix and tall fescue to 384 grazing animal days on sorghum-sudangrass. Only 3.52 inches of rain occurred from the day of planting to the last day of grazing.

[1]Extension Educator, The Ohio State University Department of Extension, Caldwell, OH (gelley.2@osu.edu).

CANOPY LAYER AND PLANT FRACTION AFFECT THE NUTRITIVE VALUE OF STOCKPILED LIMPOGRASS IN NORTH FLORIDA

M. Bernardini, J.C.B. Dubeux Jr., R.S. Reis, C.T.E. Mendes, M.D.G. Altamirano, M. Ruiz-Moreno, L. Garcia, L.D. Queiroz, K.R. Trumpp, K. Oduor, J. Portuguez, I. Bretas, N. DiLorenzo[1], and J.M.B. Vendramini[2]

Limpograss [*Hemarthria altissima* (Poir.) Stapf & C.E. Hubb.] is a stoloniferous warm-season perennial grass used by beef cattle producers in Florida. Due to its capability in maintaining its nutritive value for longer periods compared to other warm-season grasses, producers have successfully adopted limpograss in Florida. Like other warm-season grasses, limpograss presents a vertically heterogeneous canopy composition that affects forage nutritive value, particularly in taller stockpiled pasture. This project was conducted at UF IFAS NFREC in Marianna, FL from October 2023 to January 2024. Treatments included upper leaf, lower leaf, upper stem, lower stem, upper all (leaf + stem), lower all (leaf + stem), and whole plant. Treatments were allocated in a RCBD with 8 blocks. Upper leaf CP concentration decreased from 9.4% from middle October to 5.9% in early January. Upper leaf presented the greatest IVDOM concentration across all evaluations and ranged from 42 to 58% during the season. Lower leaf IVDOM followed a similar pattern as upper leaf and ranged from 47 to 53 % across evaluations. Digestibility at the upper canopy ranged from 40.1% in middle October to 28.3% in early January. Likewise, in the same season the whole plant fluctuated across evaluation and ranged from 44.6 from middle October to 32.8% in early January. Limpograss exhibits a vertical heterogeneity in nutritive value, highlighting the importance of grazing management and the need of strategic grazing or supplementation strategies to optimize animal performance.

[1]University of Florida, North Florida Research and Education Center (NFREC); [2]University of Florida, Range Cattle Research and Education Center.

BROWN MID-RIB TRAIT TRADEOFFS IN
SORGHUM-SUDANGRASS AND PEARL MILLET YIELD

A. Barnett, K.J. Simon, J.O.C. Kubesch, and M.S. Gadberry[1]

The brown midrib (BMR) trait has been used to improve the nutritive value of sorghum-sudangrass and pearl millet, but is known to affect yields. Seed cost/lb might influence seed selection; BMR seed costs twice as much as standard seed. This trial sought to combine yield data and economics for producers considering the BMR trait. A replicated field trial was conducted at the Livestock and Forestry Research Station in Batesville, AR. Pearl millet and sorghum-sudangrass with and without the BMR trait were planted 5/30/24 using an Almaco light duty grain drill at a seeding rate of 25 lb./acre. Plots were managed as either common producer practices (CPP) or best management practices (BMP), and harvested three times. The first two harvests were at a 6” stubble and the final harvest was at 2” stubble. Harvested forage mass was weighed, and sub-samples were collected for moisture and nutritive value analysis. The BMR trait did not affect harvested forage mass, nor did species selections. Plots managed according to CPP had higher forage harvests than BMP plots in the first and second harvests. Forage production costs came to a difference of <$0.005/lb between BMR and standard options of both forages. The two-fold seed cost difference seen in BMR and standard summer annual forages does not compare to the greater investments made in weed control, soil fertility, and harvest management. These results suggest farmers should prioritize harvest management regardless of species selection or BMR status, especially on sites with high weed competition.

[1]Fulton County Agent, Univ. of Arkansas System Div. of Agric., Coop. Ext. Service, Salem, AR (abarnett@uada.edu); Forages Instructor, Univ. of Arkansas System Div. of Agric., Coop. Ext. Service, Little Rock, AR (ksimon@uada.edu); Forage Specialist, Univ. of Arkansas System Div. of Agric., Coop. Ext. Service, Little Rock, AR (jkubesch@uada.edu); Dir. Livestock and Forestry Research Station, Univ. of Arkansas, Batesville, AR (sgadberry@uada.edu).

AGRONOMIC AND NUTRITIVE VALUE RESPONSES OF DUAL-PURPOSE OAT AND WHEAT CULTIVARS UNDER CONTRASTING HARVEST STRATEGIES

Daniela Rodriguez[1], Carlos Garcia, Jacob Adkison, Katie Seavey, Luana Zagato, Alexa Cornett, Leandro Vieira, Alex Coleman, and Liliane Silva[1]

Cool-season forages provide forage production for fall and winter months when perennial warm-season grasses are dormant. This study evaluated agronomic responses and structural characteristics of oat *(Avena sativa)* and wheat *(Triticum aestivum)* cultivars managed under simulated grazing or grain production at two South Carolina locations. Four oat (Horizon 306, Horizon 578, Brooks, and Horizon 214) and one wheat (AGS 4023, control) cultivars were used. Two harvests were conducted in the simulated grazing treatment in February and March of 2024. Forage mass was greater in March than February for all cultivars at both locations averaging 3,723 vs. 1,556 and 2,790 vs. 1,206 kg DM/ha, respectively. At second harvest, AGS4023 had greater neutral detergent fiber concentration than the oat varieties. When averaging both locations, AGS 4023 wheat and SCLA214 oat tended to exhibit greater crude protein (CP) (P=0.066). Also, CP tended to be greater in all varieties planted in Edisto (P=0.0843). However, there were no significant differences in CP between varieties and REC. Oat still produced substantial forage mass and CP concentration, making them effective for grazing during the cool-season. Further research is needed to adapt grazing management strategies for use of dual-purpose practices for the Southeast.

[1]Master Student, Clemson Univ., Blackville, SC (yulietr@clemson.edu); Postdoctoral Fellow, Clemson Univ., Blackville, SC (cvelaga@clemson.edu); Field Technician, Clemson Univ., Blackville, SC (dadkiso@clemson.edu); Master Student, Clemson Univ., Blackville, SC (kseavey@clemson.edu); Ph.D. student, Clemson Univ., Blackville, SC(lquirin@g.clemson.edu); Ph.D. student, Clemson Univ., Clemson, SC(aharve6@g.clemson.edu); Postdoctoral Fellow, Clemson Univ., Blackville, SC (lvieira@clemson.edu); Extension Specialist, Clemson Univ., Clemson, SC (amcolem@clemson.edu); Assistant Professor, Clemson Univ.y, Blackville, SC (lseveri@clemson.edu).

MYCOTOXIN DETECTION METHODS AT CHEMICAL AND MOLECULAR LEVELS

Haihua Wang, Kaile Zhang, Benjamin Reimer, Robbie Jones, Vijay Verma, Brittany Justesen, Joeseph Walter, Ann Blount, and Hui-Ling Liao

The development of a mycotoxin detection technique, targeting both chemical content and gene expression of the associated microorganisms/mycotoxin-producing pathways, offers the potential to predict mycotoxin production in forage and hay samples. This advancement would significantly benefit pasture health and management, ensuring safer feed for livestock. Currently, the primary methods for detecting mycotoxins in food and pasture involve high-performance liquid chromatography (HPLC) and liquid chromatography-mass spectrometry (LC-MS/HPLC). While these techniques are highly effective in quantifying mycotoxins, they are limited to the detection of chemical components. Consequently, it is necessary to apply molecular tools that can enhance the predictive capabilities in the inspection of mycotoxins by identifying molecular mycotoxin-producing pathways. Here, we have developed a comprehensive mycotoxin diagnostic approach that incorporates molecular detection techniques. This includes amplicon sequencing to evaluate the abundance of mycotoxin-producing microorganisms and real-time quantitative PCR (qPCR) to sensitively detect the expression of genes associated with mycotoxin production. By integrating these molecular methods with LC-MS/HPLC quantification, our approach not only detects the current levels of mycotoxins in samples but also enables us for future predictive capabilities. This combined strategy enhances the potential for early intervention in the feed production chain, reducing the risk of mycotoxin contamination and improving overall livestock health and productivity.

IMPACT OF DRYING TEMPERATURE ON ISOFLAVONE CONCENTRATIONS IN RED CLOVER

J.L. Stephens, C.D. Teutsch, and S.R. Smith[1]

One strategy for mitigation of negative symptoms of tall fescue toxicosis in cattle involves the intake of dried red clover (*Trifolium pratense* L.) via loose mineral amendment. While evidence supporting this method exists, there are no red clover-amended loose mineral mixes available for purchase. To develop such a product, scaling up red clover production is required. Red clover leaves have the highest levels of toxicosis-mitigating isoflavone compounds, especially Biochanin A, out of all the plant's tissues. To avoid losses due to shatter from field drying, it has been proposed to bale red clover at higher moisture levels. The plant material will then be dried down in a rotary drum dryer prior to granulation. Rotary drum dryers use temperatures more than 180 degrees Fahrenheit. It is unknown if red clover isoflavones are affected by temperatures greater than 160 degrees Fahrenheit. In this study, samples of red clover were oven-dried at 140, 160, 180, 200, and 220 degrees Fahrenheit and compared to freeze-dried samples for a control group. Following drying, samples were ground, then isoflavones were extracted and analyzed via UHPLC-MS/MS. While freeze-drying resulted in greater levels of isoflavones than any oven-dried samples, no differences were observed between any of the oven-dried samples. Therefore, these initial findings are promising for scaling up red clover production with the end goal of producing a red clover-amended livestock mineral.

[1]Master's Student, Univ. of Kentucky, Dept. of Plant and Soil Sciences, Univ. of Kentucky Research and Education Center, Princeton, KY (jonathan.stephens@uky.edu); Extension Associate Professor, Univ. of Kentucky, Dept. of Plant and Soil Sciences, Univ. of Kentucky Research and Education Center, Princeton, KY (chris.teutsch@uky.edu); Extension Professor, Univ. of Kentucky, Dept. of Plant and Soil Sciences, Agricultural Science Center North, Lexington, KY (raysmith1@uky.edu).

AGRONOMIC AND STRUCTURAL RESPONSES OF ANNUAL WARM-SEASON FORAGE MIXTURES MANAGED UNDER CONTRASTING HARVEST MANAGEMENT

L. Silva, C. Garcia, D. Adkison, K. Seavey, L. Vieira, D. Rodriguez, and L. Zagato[1]

Warm-season annual grass, legume, and brassica forages are viable options to supply biomass to support livestock operations in the southeast region. There is need to better understand agronomic responses of monoculture, binary, and mixtures of warm-season annual forage under simulated grazing or baleage production. The study was conducted at the Edisto Research and Education Center in Blackville, South Carolina. Treatments consisted of twelve forage treatments including eight monocultures, and four binary or multi-species (3+ species) mixtures planted on June 6th with harvests on July 7th, Aug 2nd and Sept 5th for the grazing treatment and on Aug 23rd for the baleage treatment. Forage mass and nutritive value were determined. Manual botanical composition was conducted, and canopy height was measured prior to each harvest event. The stubble height used was 16”. Weed control was superior on the second harvest (August) for all treatment and doubled for some of the forage mixtures. In the mixture treatments, higher legume percentages were observed at the second harvest period. Limited rainfall in early summer slowed forage establishment and impacted initial growth. The preliminary results corroborate that forage mixtures support complemental forage distribution and quality but these warm-season annual require proper management to avoid them becoming overly mature quickly with impacts seasonal production and decreases forage quality.

[1] Assistant Professor, Clemson University, Blackville, SC (lseveri@clemson.edu); Postdoctoral Fellow, Clemson University, Blackville, SC (cvelaga@clemson.edu); Field Technician, Clemson University, Blackville, SC (dadkiso@clemson.edu); Master Student, Clemson University, Blackville, SC (kseavey@clemson.edu); Postdoctoral Fellow, Clemson University, Blackville, SC (lvieira@clemson.edu); Master Student, Clemson University, Blackville, SC (yulietr@clemson.edu); Ph.D. student, Clemson University, Blackville, SC (lquirin@g.clemson.edu).

UPDATING NUTRITIVE VALUE EXPECTATIONS FOR DIFFERENT SUMMER ANNUAL FORAGE MANAGEMENT STYLES

M.L. Mobley, K.J. Simon, J.O.C. Kubesch, and M.S. Gadberry[1]

Arkansas Extension publications report average summer forage quality based on forage test results produced in the state. However, characteristics can vary with different management styles. Frequent fall droughts also raise concerns about nitrate poisoning in livestock. This experiment was designed to generate the nutritive value and nitrate level of sorghum-sudangrass and pearl millet with and without the brown midrib trait under two different harvest management regimes. The trial was conducted at the Livestock and Forestry Research Station in Batesville, AR. Plots were no-till planted on 5/30/24 at 25 lb./acre. Plots were harvested three times. The first two harvests were cut to 6" stubble and the final harvest was at 2" stubble. Forage mass was weighed, and sub-samples were collected for moisture and nutritive value analysis. Management style was the main factor affecting nutritive value. Plots managed with best management practices (BMP) had higher crude protein (CP) content and total digestible nutrients (TDN) than those managed with common producer practices (CPP). Average nutritive values for sorghum-sudangrass were slightly lower and CP of pearl millet was higher than those reported in FSA2032. Nitrate levels were within safe guidelines across harvests. This trial quantifies nitrate risk in summer annual grasses with and without the BMR trait, under different harvest regimes. This trial will be repeated to establish clear expectations for nutritive value and nitrate levels.

[1]Independence Co. Staff Chair, University of Arkansas System Division of Agriculture, Cooperative Extension Service, Batesville, AR (mmobley@uada.edu); Forages Instructor, University of Arkansas System Division of Agriculture, Cooperative Extension Service, Little Rock, AR (ksimon@uada.edu). Forage Specialist, University of Arkansas System Division of Agriculture, Cooperative Extension Service, Little Rock, AR (jkubesch@uada.edu); Director of the Livestock and Forestry Research Station, University of Arkansas System Division of Agriculture, Batesville, AR (sgadberry@uada.edu).

IMPACT OF N FERTILIZATION AND INOCULATION ON NUTRITIVE VALUE AND NITRATE LEVELS OF SORGHUM-SUDANGRASS BALEAGE

R. Bustabad[1], K.M. Mason[1], B.C. Pedreira[2], J.L.Z. Ivey[1], D. McIntosh[2], and M. Fancher[2]

Livestock producers often use warm-season grasses, such as sorghum-sudangrass (*Sorghum x drummondii*), to create additional grazing days or conserved forage in the summer. Under stressful conditions, especially drought, sorghum-sudangrass may accumulate nitrates (NO_3^-), resulting in potential nitrate poisoning when consumed. Some evidence shows that ensiling can increase nitrate degradation in contrast to common hay production. The objective of this study is to determine the effect of nitrogen (N) fertilization on nitrate levels and nutrient concentration of sorghum-sudangrass and to estimate nitrate reduction following ensiling with or without bacterial inoculant. 'Greengrazer' sorghum-sudangrass plots were planted in May 2024. Nitrogen rate treatments were 0, 45, 90, 135, and 180 lb N/ac. In August, harvested samples were weighed and analyzed for nutritive value. On average, NIR analysis results indicated 7.8, 9.0, 11.3, 15.8, and 16.7% crude protein (CP) concentrations with forage masses of 4,468, 4,966, 6,128, 7,270, and 5,867 lb/ac, respectively for each N rate treatment. Samples from five groups of forage with varying nitrate levels were wilted to 55% moisture. Half were inoculated with Promote® Forage-Mate® EBL, then all samples were placed into vacuum-sealed bags for ensiling. Following a 6-week fermentation, NIR analysis indicated 7.3, 8.9, 10.3, 15.0, and 16.2% CP concentration, respectively, for each treatment group. Results will be used to inform producers of preservation options for high-nitrate forage.

[1]Graduate Assistant (rbustaba@vols.utk.edu); Assistant Professor/Extension Specialist; Associate Professor/Extension Specialist, Animal Science, University of Tennessee, Knoxville, TN. [2]Associate Professor/Extension Specialist; Coordinator/Researcher; Research Assistant, Plant Sciences, University of Tennessee, Knoxville, TN.

COMPARISON OF COMMERCIALLY AVAILABLE HAY MOISTURE TESTERS

W. Peck, K.A. Trimble, and C.D. Teutsch[1]

Abstract

Hay should be baled between 16 and 20% moisture depending on bale size and type. There are a number of methods for determining hay moisture at baling including feel, microwave drying, forced air ovens, Koster type dryers, and electrical hay probes. While oven and microwave drying provide the most accurate estimations of hay moisture their use in the field is not practical. Hay probes that use the principle of electrical resistance are the least accurate but most practical to use. There are a number of commercially available hay probes. The objective of this study was to evaluate and compare the accuracy of three different hay probes for estimating hay moisture against a conventional forced air oven. The experimental design was a random complete block with 12 replications. At each sample site (four per bale), hay moisture was determined with each hay probe and a core was taken for oven drying. There was not a significant difference between the Delmhorst (16.9%) and AgraTronix (16.7%) or the Draminski (15.6) and oven dried (15.3%). However, all the probes were within approximately 10% of the oven dried sample indicating relatively good performance under the conditions of this experiment. Our data indicate that all three hay probes could provide reasonable estimates of the hay moisture at baling.

Introduction: Baling hay that is too high in moisture can lead to mold formation, spontaneous heating and in some cases combustion, loss of dry matter, and decreases in nutritive value (Collins and Coblentz, 2007). For safe storage, hay moisture at baling should be between 16 and 20% depending on bale size and type (Collins and Coblentz, 2007). There are a number of methods for determining hay moisture at baling including microwave drying, forced air ovens, Koster type dryers, and electrical hay probes (Shewmaker and Thaemert, 2004). While oven and microwave drying provide the most accurate estimations of hay moisture at baling their use in the field is not practical. Hay probes that use the principle of electrical resistance are the least accurate but most practical to use in the field (Shewmaker and Thaemert, 2004). These probes can provide an almost instantaneous estimates of hay moisture. There are a number of commercially available hay probes. The objective of this study was to evaluate and compare the accuracy of three different hay probes for estimating hay moisture against a conventional forced air oven.

Materials and Methods: The experimental design was a random complete block with 12 replications. Twelve bales were randomly selected and four locations on each bale were tested (Fig. 1). Bales served as replications. Treatments were three electronic hay probes: 1) Delmhorst, F-2000 (Towaco, NJ), 2) Draminski, High Moisture (Sząbruk, Poland), Agratronix, HST-1 (Streetsboro, OH), and a forced-air drying oven (Grieve SC-400, Round Lake, IL) (Fig. 1). A description of the hay probes can be found in Table 1. Some bales were located outside, and some bales were located inside a hay barn. At each sample site (four per bale), hay moisture was

[1] W. Peck, undergraduate, University of Kentucky, Lexington, KY; K.A. Trimble, undergraduate, Murray State university, Murray, KY; and C.D. Teutsch, Forage Specialist, UK Research and Education Center, Princeton, KY.

determined with each hay probe and a core was taken for oven drying (Fig. 2). The hay probes were randomized at each sample site to one of the three potential locations (Fig. 2). Testers were then inserted to a depth of 18-inches and allowed to stabilize before readings were recorded. A core was then taken with a Star Uni-Forage Sampler with spiral assist (Star Quality Samplers, Inc, Irricana, AB, Canada) in the center of each sample site to a depth of 18-inches (Fig. 2). The cored samples were dried for 3-4 days at 131°F. Tightly bond moisture was determined by using near infrared spectroscopy (McIntosh et al., 2022). The total dry matter was calculated using the following formula: Total DM = Partial DM x (Lab DM/100) (Undersander et al., 1993). Data was then analyzed using the General Model Procedure from SAS. Means were separated using Fisher's Protected Least Significant Difference ($\alpha = 0.05$).

Results and Discussion: There was not a significant difference between the Delmhorst and AgraTronix or the Draminski and oven dried (Fig. 3). Both the Draminski and the oven dried sample had lower moisture levels (Fig. 3). Descriptive statistics are shown in Table 2. The oven dried sample had the smallest range, approximately one-half that of the Delmhorst and Draminski and smallest standard deviation (2.3 versus 3.86) (Table 2). The hay probes were likely more sensitive to small areas of wet forage in the bales since the distance between the sensors reading the electrical resistance is relatively small (1/2 to 1-inch) (Fig. 4). In contrast, the moisture is determined over the entire length of the 18-inch hay core in the oven drying method. The probes were also scored on ease of use by the operators with the Agratronix scoring the highest (Table 1).

Summary and Implications: The Draminski hay probe produced results closest to the oven drying method indicating that it was the most accurate of the three probes evaluated. An additional benefit of this probe is its ability to measure bale temperature, a feature that could help be useful in barn fire prevention. However, all the probes were within approximately 10% of the oven dried sample indicating relatively good performance under the conditions of this experiment. Our data indicate that all three hay probes could provide reasonable estimates of the hay moisture at baling. Considering ease of use and price (Table 1) could help producers choose the hay probe best suited for their operations.

Literature Cited:

Collins, M. and W.K Coblentz. 2007. Post-harvest physiology. In Forages, The Science of Grassland Agriculture, Volume II, 6th Edition. Blackwell Publishing, Ames, IA.

McIntosh, D., B. Anerson-Husmoen, R. Kern-Lunbery, P. Goldblatt, R. Lemus, T. Griggs, L. Bauman, S. Boone, G. Shewmaker, and C. Teutsch. 2022. Guidelines for the Optimal Use of NIRSC Forage and Feed Calibrations in Membership Laboratories (Second Edition). The UT Press, Knoxville, TN.

Shewmaker, G.E. and R. Thaemert. 2024. Measuring moisture in hay. In the Proceedings of the 34th California Alfalfa Symposium held in Monterey on December 17-19, 2004. Available online at https://alfalfasymposium.ucdavis.edu/+symposium/proceedings/?yr=2004.

Undersander, D., D.R. Mertens, and N. Thiex. 1993. Forage Analyses Procedures. NFTA, Stuart, FL. Available online at https://www.foragetesting.org/_files/ugd/24f64f_4502c4143ec34a6ba1c4940c6147fd07.pdf.

Table 1. Description of hay probes used in this experiment including the model, manufacturer, moisture range, price, temperature probe, and ease of use rating.

Probe Model	Manufacturer	Moisture Range*	Price†	Ease of Use‡
F-2000	Delmhorst Instument Co., Towaco, NJ	6-40%	$400	3/5
High Moisture	Draminski, Sząbruk, Poland	10-80%	$375	3/5
HST-1	Agratronix, Streetsboro, OH	8-72%	$280	4/5

*Moisture range was provided by manufacturers.

†Established form survey of on-line offerings.

‡Ease of use rate was subjective rating that took into account assembly required, instruction manual clarity, and ease of operation.

Table 2. Descriptive statistics for moisture meters and drying oven.

Statistic	Agratronix	Delmhorst	Draminski	Oven
	----------	---------- % ----------	----------	----------
Average	16.7	16.9	15.6	15.3
Median	16.2	16.7	14.8	15.1
Min	11.8	8	10.0	8.9
Max	27.6	32	32.2	21.3
Range	15.8	24	22.3	12.4
Standard Deviation	3.2	4.7	3.7	2.29
Observations (n)	48	48	48	48

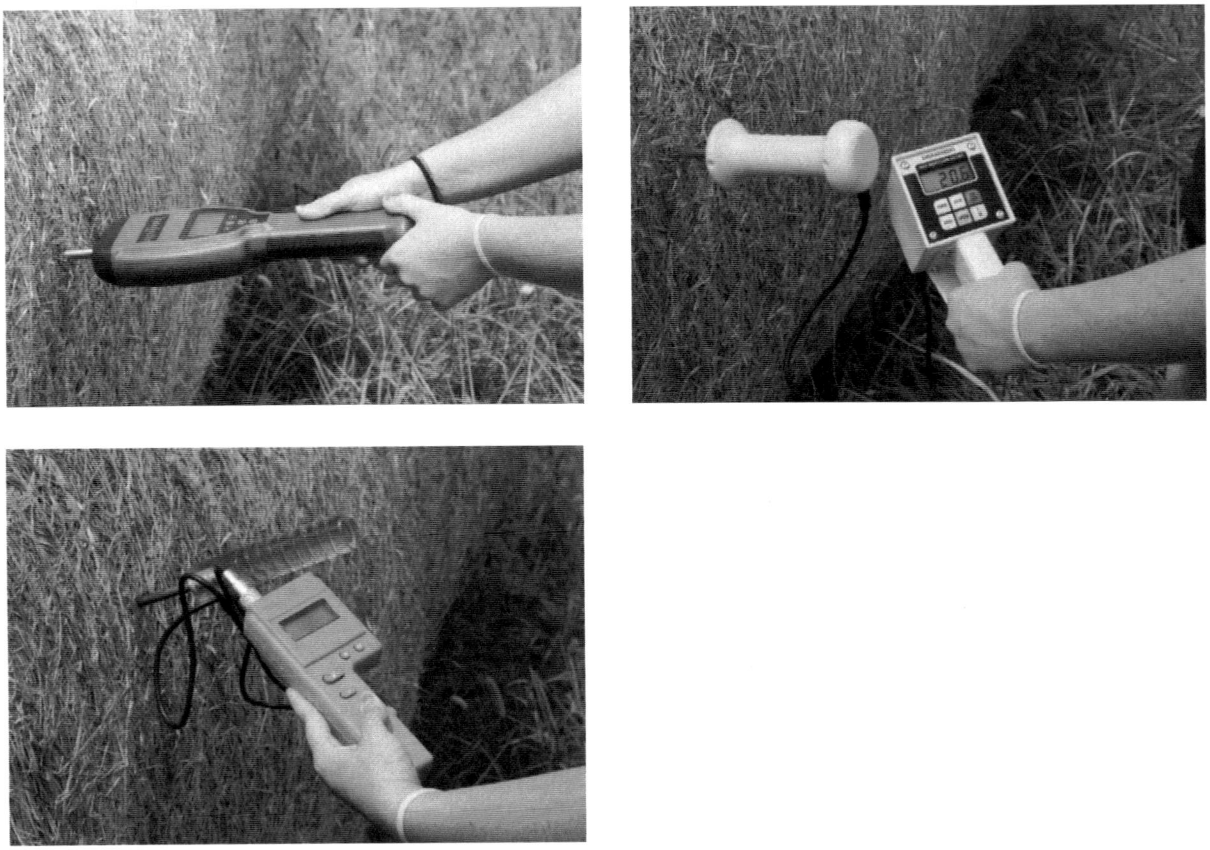

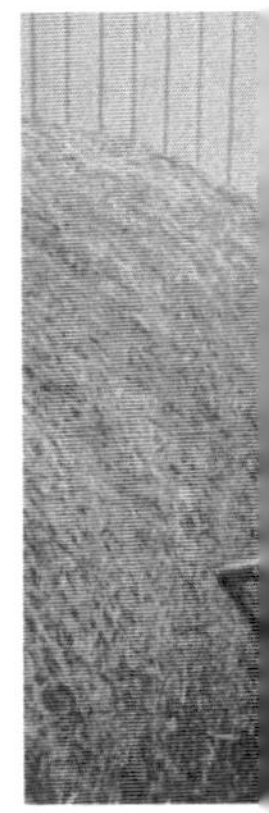

Figure 1. Hay moisture probes evaluated included Agratronix HST-1 (top left), Draminski High Moisture (top right), and Delmhorst FC-2000 (bottom left). Hay probes were compared to a core sample that was dried in a conventional forced-air oven.

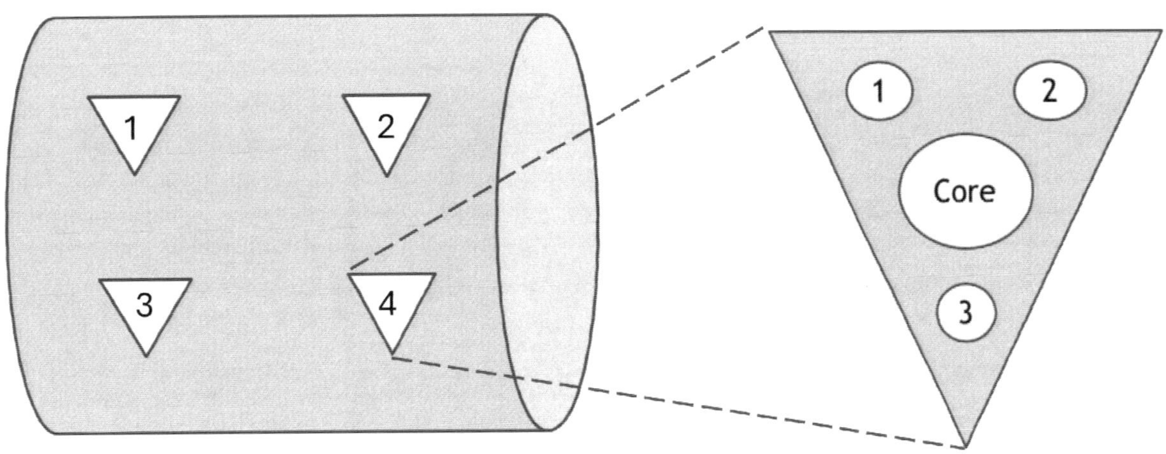

Figure 2. Four samples were collected from each bale (left). Hay probes were randomized at each sample site to one of the three potential locations (right). A core sample was collected in the center of each sample site (right).

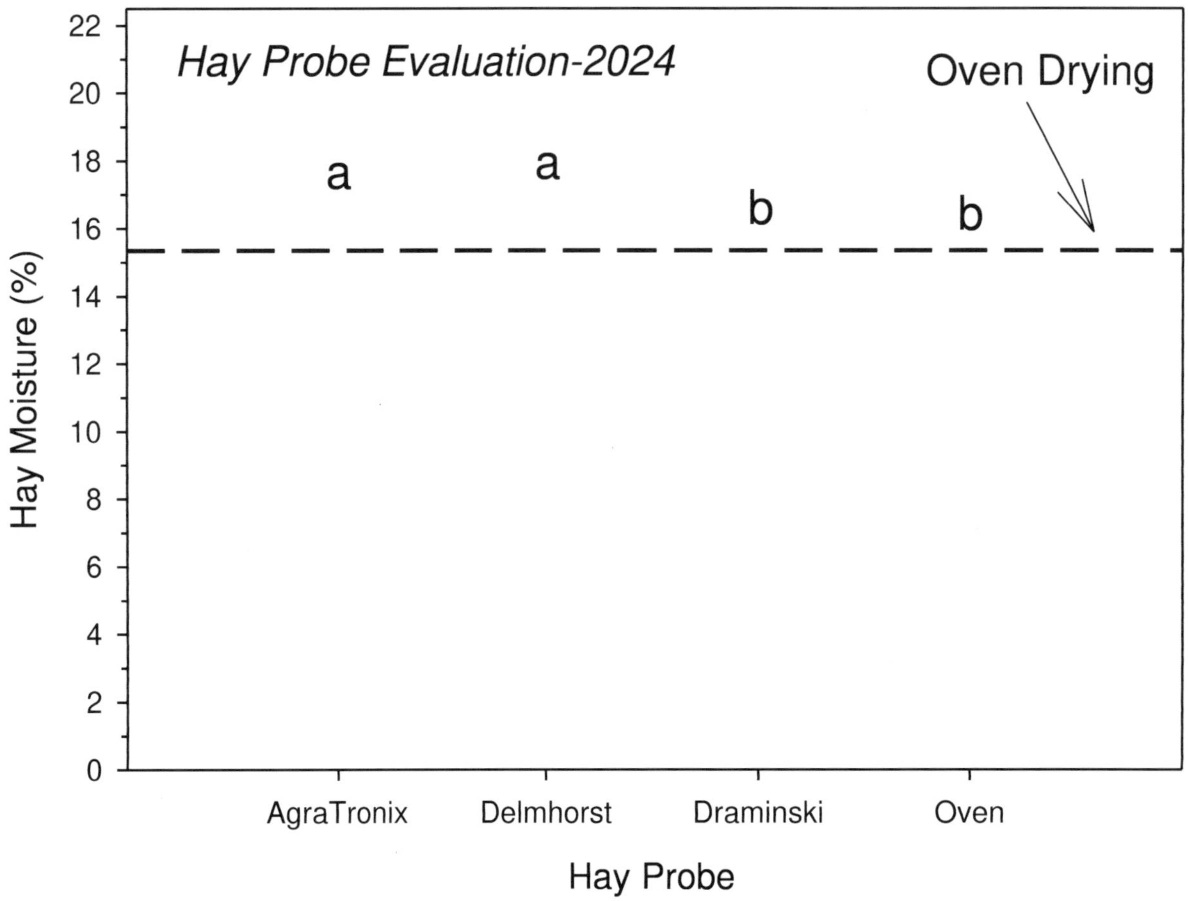

Figure 3. Hay moisture as estimated by the three hay probes and oven drying. Bars with the same letter are not different according to Fisher's Protected Least Significant Difference, α = 0.05.

Figure 4. Electrical resistance is read across the insulated portion of the hay probe. In the case of the probes used in this experiment this distance ranged from ½ to 1-inch.

FORAGE ANALYSIS BY NEAR INFRARED SPECTROSCOPY (NIRS) VS. WET CHEMISTRY: A COMPARATIVE STUDY BY NIRS CONSORTIUM

U.K. Saha[1] and R. Kern-Lunbery[2]

Near infrared spectroscopy (NIRS) is an advantageous method for forage analysis. It is nondestructive, less labor intensive, rapid, environmentally friendly, yet accurate and precise results. However, there is a misconception that the NIRS method has a 'poor accuracy'. Therefore, in this study, we evaluated the accuracy and precision of dry matter (DM), crude protein (CP), acid detergent fiber (ADF), and neutral detergent fiber (NDF) of 33 National Forage Testing Association (NFTA) proficiency test (PT) alfalfa hay samples analyzed by NIRS in 7 NIRS Forage and Feed Testing Consortium (NIRSC) member laboratories. The reference method averages (RMA), used to evaluate the NIRS results, were based on the wet chemistry results reported by numerous laboratories participating in the corresponding NFTA proficiency testing rounds. Thus, this study is a robust comparison of NIRS results with the corresponding wet chemistry results, which is still a "gold standard" to many nutritionists. These results demonstrate that when NIRS calibrations are developed using good science and applied properly, both intra-laboratory and inter-laboratory precision of NIRS methods are superior to wet chemistry methods in forage nutritional analysis. Further, NIRS results are as accurate as wet chemistry.

[1]Agricultural and Environmental Services Laboratories, University of Georgia, [2]WARD Laboratories, Inc.

TIMELY BOTANICAL COMPOSITION ESTIMATION IN GRASS-LEGUME PASTURES TO GUIDE THE GRAZING MANAGEMENT

Igor L. Bretas[1], Zhou Tang[2], Jose C.B. Dubeux Jr.[1], Chang Zhao[2], Kevin R. Trumpp[1], Javier P. Acuna[1], Cleber H. L. de Souza[1], and Iam M. Small[3]

Estimating botanical composition in grass-legume pastures is crucial for maintaining a stable and harmonic sward. To achieve this balance, grazing management should aim for a legume proportion of 20-45% of the forage dry matter. Conventional methods for tracking botanical composition are often time-consuming, labor-intensive, and inaccurate. A study at the University of Florida (NFREC) used drone images and machine learning to classify grass and legumes in mixed pastures and estimate their proportions. Data were collected from nine paddocks under three treatments: unfertilized bahiagrass (*Paspalum notatum* Flüggé), fertilized bahiagrass, and bahiagrass-rhizoma peanut (*Arachis glabrata* Benth.) pastures. Two multispectral images (taken in early and late summer) were captured using a drone at 30m altitude. The input variables included individual band values and five vegetation indices: NDVI, VARI, ExG, NDRE, and SAVI. Grass (n=360) and legumes (n=216) were labeled as ground truth data using a Real-Time Kinematic system with an average accuracy of 1.5 cm. Approximately 80% of data points were used for model training with 5-fold cross-validation, while 20% were reserved for testing. The random forest algorithm achieved an overall prediction accuracy of 89%. The predicted legume cover in mixed pastures averaged 19% in early summer but dropped in late summer (6%). Multispectral data allows the botanical composition estimation in mixed pastures, though senescent material and bare soil present challenges later in the growing season.

[1]University of Florida, North Florida Research and Education Center (NFREC), Marianna, FL, USA; [2]University of Florida, Agronomy Department, Gainesville, FL, USA; [3]University of Florida, North Florida Research and Education Center (NFREC), Quincy, FL, USA.

AUTHOR INDEX